PreTest®
Step 1 Simulated Examination

PreTest®
Step 1 Simulated Examination

8th Edition

Editor
John R. Thornborough, Ph.D.
Mount Sinai School of Medicine
New York, New York

Contributors
G. Thompson Burke, Ph.D.
Steven H. Dikman, M.D.
Joseph S. Eisenman, Ph.D.
John P. Morgan, M.D.
Martin S. Nachbar, M.D.
Todd R. Olson, Ph.D.

McGraw-Hill
Health Professions Division/PreTest® Series

New York   St. Louis   San Francisco   Auckland
Bogotá   Caracas   Lisbon   London   Madrid
Mexico City   Milan   Montreal   New Delhi
San Juan   Singapore   Sydney   Tokyo   Toronto

McGraw-Hill
A Division of The McGraw-Hill Companies

PreTest® Step 1 Simulated Examination, Eighth Edition

1 2 3 4 5 6 7 8 9 MALMAL 9 9 8 7 6

ISBN 0-07-052020-8

This book was set in Times Roman by ILOC, Inc.
The editors were Gail Gavert and Deborah L. Harvey.
The production supervisors were Anna Lieggi and
Diana Porter Jones.
Malloy Lithographers, Inc., was printer and binder.

This book is printed on 85% recycled, 10% postconsumer waste, acid-free paper.

Library of Congress Cataloging-in-Publication Data

PreTest step 1 simulated exam. -- 8th ed. / editor, John R.
     Thornborough ; contributors, G. Thompson Burke . . . [et al.]
          p.   cm.
     Rev. ed. of: PreTest National Medical Board examination
comprehensive part I. 7th ed. © 1991.
     Includes bibliographical references.
     ISBN 0-07-052020-8
     1. National Board of Medical Examiners--Examinations--Study guides.
2. Medicine--Examinations, questions, etc.   I. Thornborough, John
R. II. Burke, G. Thompson.   III. PreTest National Medical Board
examination comprehensive part I.
     [DNLM: 1. Medicine--examination questions.   W 18.2 P9424 1995]
R834.5.P725   1995
610'.76--dc20
DNLM/DLC   95-23413
for Library of Congress

Contents

Introduction

PreTest® Step 1 Simulated Examination has been designed to help students as well as physicians assess their knowledge of the basic medical sciences and prepare for Step 1 of the United States Medical Licensing Examination (USMLE).

The overall format of this simulated examination—including the number, item type, and degree of difficulty of the questions; the scoring; the use of illustrative materials; and the time limits imposed—parallels the latest format of the actual Step 1 examination. Because the USMLE examination necessarily varies in content from year to year, PreTest does not attempt to duplicate exactly the Step 1 examination; rather, it presents questions on topics and problems, based on both fundamental concepts and recent advances that are of key importance in the following basic science areas: anatomy, behavioral sciences, biochemistry, cell biology, genetics, histology, microbiology, neuroscience, pathology, pharmacology, and physiology.

Approximately 40 percent of the questions are new and written specifically for this examination. The remainder of the questions were carefully edited, evaluated, reviewed, and updated for this examination from previous PreTest exams and books.

PreTest's computerized evaluation service, included free of charge with this test book, generates a score for the total test as well as scores for each of the above-mentioned subject areas. (A passing score for the USMLE examination is based *only* on performance on the total test, i.e., there is no minimum passing score required for an individual subject area.) PreTest evaluations are designed to highlight areas of expertise and identify areas of weakness in the basic sciences.

Each question is accompanied with a brief answer/explanation and a reference. The references and bibliography that are included with the package should be used as a guide for a more extensive review of the subject areas reflecting a particular weakness.

PreTest® Step 1 Simulated Examination is approximately half the length of the actual Step 1 examination. It is divided into two sections of 210 questions each (total number of questions equals 420) and requires one full day of testing. The USMLE Step 1 requires two full days of testing. Both examinations present the material in interdisciplinary form–that is, all sections of the examination contain questions concerning all of the basic sciences without identifying questions as belonging to a given area. Moreover, many questions are "cross disciplinary" in that they contain material from more than one area.

Directions for Use

TYPES OF QUESTIONS

There are two basic types of questions in this examination.

Type 1 – **One best answer completion.** A question or incomplete statement is followed by 3 to 26 answers or completions. You are to select the single **best** choice.

Type 2 – **Matching questions.** A group of 3 to 26 lettered answers or a lettered diagram is followed by a list of numbered phrases or statements. For each numbered phrase or statement, you are to select the one answer that is most closely related to it. Each answer may be selected once, more than once, or not at all.

Each type of question has only **one best answer**. You should realize that sometimes no clear-cut choice exists in a situation, just as is the case in the actual practice of medicine. Therefore, although several choices may appear appealing, the less applicable ones must be rejected in favor of the **single best** answer. Make sure you do not mark more than one space per question.

TIME ALLOTMENT

The USMLE Step 1 examination allows approximately 50 to 60 seconds per question—the same amount of time allotted in most medical tests. In accordance with this time allotment, there is a three hour limit for each of the two sections of this simulated examination. Each section of this examination should be treated as a separate test. Thus, you should plan to take each section of this PreTest at a time when you will be undisturbed for a minimum of three hours. Choose a quiet, well-lit location, free from distractions. The area should be clear of books, papers, or other materials, and a clock or watch should be easily visible. Have three well-sharpened **No. 2** pencils and an eraser available. If a section is completed within the allotted time, any remaining time may be used to review questions in that section. Extra time from one section should **not** be spent in answering questions in another section.

MARKING THE ANSWER SHEETS

The answer sheets are located at the back of this test book. *Carefully* tear along the perforated lines to remove them. Try to keep them flat and unmarked, especially if you wish to send them to McGraw-Hill for free evaluation and grading.

Using a **No. 2** pencil only, enter your name, mailing address, and your five-digit identification number (which is stamped on the front of your answer sheets) in the appropriate boxes and blacken the corresponding positions in the grids on your answer sheets, as shown above. *It is very important that you enter your name and mailing address clearly and accurately on the answer sheet.* The optical scanner will read this information to generate a mailing label used for the return of your scores. If you need more room than is available, print your name and address clearly on a Post-it™ note or a separate sheet of paper and include it with your answer sheets.

Each section of this examination has a separate sheet for entering your answers. **Be sure to enter your five-digit identification code in the appropriate boxes of every page of the answer sheets.** For each question in this examination, fill in only one space on your answer sheet next to the appropriate question number. Be sure your marks fill the space *within the correct box*, but do not stray *outside the box*. If you erase, be certain that you make a clean erasure.

If your answer sheet is not filled out according to these guidelines, a scanner will not be able to read your answers correctly. As a result, either your sheet will be returned to you unscored or your computer scoring summary will carry the following message: "Because of your incomplete erasures or faulty recording of answers, there may be inaccuracies in this scoring."

IF YOU WISH TO HAVE YOUR ANSWER SHEETS EVALUATED

When you have completed the examination, you can check your answers against the answer/explanation material provided at the end of the two examinations. Or you can fold the sheets carefully at the "fold" marks indicated, place them in a legal-size envelope, and mail them to:

McGraw-Hill/PreTest
HPD 28
1221 Avenue of the Americas
New York, NY 10020
Attn: Test Marking.

The answer sheets must have *all* questions answered and the identification codes and your name and address entered. All answer sheets must be returned together.

SCORING

McGraw-Hill/PreTest determines the score for this test on the basis of the number of questions answered correctly; the same system of scoring is used by the USMLE.

Because there is no penalty for wrong answers, you should guess at an answer to every question for which you do not know the answer. **PreTest examinations are computer scored every two weeks throughout the year.** Please allow ample time for your test results to be returned to you.

INTERPRETING THE SCORES

A raw score, a percentage correct, a percentile score, and a standard score are reported for each of the basic medical sciences listed in the introduction, as well as for the examination as a whole. The **raw score** is the number of questions answered correctly. (On the actual USMLE examination, a minimum passing score usually results from correctly answering 55 to 60 percent of the total number of questions in the test.) The **percentage correct** is calculated by dividing the raw score by the number of items in the examination (420). All examinees first are ranked according to their raw scores. Ranking is then expressed relative to the percentage of examinees who received a lower score, that is, as a **percentile score**. For example, an examinee answering 223 questions correctly on the examination would have a raw score of 223, and a percentage correct of 53. If 66 percent of the examinees answered fewer than 223 questions correctly, then this examinee's percentile score would be 66.

Because many licensing exams use a standard score system, the percentile score is then converted to a **standard score**. The mean score is set at 200 and one standard deviation equals 20. Traditionally, the minimum passing score of the actual licensing examination has been 1.2 standard deviations below the average standard score of the candidate reference group, that is, 176. This results in a failure rate of approximately 12 to 15 percent.

It should be noted that the percentile and standard scores reported for the total test are not the average of these scores as determined for the individual subject areas. This is a consequence of the fact that the percentile and standard scores for the entire test are generated directly from the raw score for the entire test.

Too detailed a comparison of one's ranking with another's on the basis of percentile score is not warranted. A conservative approach would not attach too much significance to scores that differ by less than 10 percentile points. As with percentile scores, not much significance should be attached to standard scores that differ by less than 10 points.

It is very important to realize that your PreTest scores may not accurately predict the scores you will attain on the USMLE Step 1 examination. Because your PreTest scores reflect your **present** level of knowledge, your standing relative to the others who have taken this exam may change between now and the time you take the official examination–especially if, in the interim, you study those areas in which your PreTest results indicate weakness. In addition, it must be remembered that PreTest scores reflect a ranking only among examinees taking this PreTest, **not** among examinees taking the official licensing examinations.

CONCLUSION

We believe that the *PreTest® Step 1 Simulated Examination*, together with the computer scoring and references for each question, provides a stimulating educational program. We hope you will find this examination valuable in strengthening your knowledge in the basic medical sciences.

PreTest®
Step 1 Simulated Examination

Book A

Time: Three hours

Number of items: 210

Book A

DIRECTIONS: The questions below consist of lettered headings followed by a set of numbered items. For each numbered item select the **one** heading with which it is **most** closely associated. Each lettered heading may be used **once, more than once, or not at all.**

Questions 1-3: Match each of the descriptions below with the appropriate region of the kidney.

    (A) Afferent arteriole

    (B) Ascending limb of the loop of Henle

    (C) Collecting duct

    (D) Descending limb of the loop of Henle

    (E) Glomerulus

    (F) Macula densa

    (G) Proximal tubule

    (H) Vasa rectae

1. Site at which the isotonic reabsorption of sodium occurs

2. Site at which the permeability to water varies with plasma osmolarity

3. Site of the active transport system that makes it possible for the kidneys to excrete a concentrated urine

Questions 4-6: Match each statement with the correct drug.

    (A) Aldosterone

    (B) Clomiphene

    (C) Diazoxide

    (D) Fludrocortisone

    (E) Isophane insulin

    (F) Methimazole

    (G) Ethinyl estradiol

    (H) Norethindrone

    (I) Norethynodrel

    (J) Propylthiouracil

    (K) Salicylates

    (L) Spironolactone

    (M) Tamoxifen

    (N) Triamcinolone

4. This drug promotes the synthesis of factors II, VII, IX, and X and may interfere with the effect of warfarin or may result in thromboembolic phenomena

5. The therapeutic effect of this drug is reduced by glucocorticoids, dextrothyroxine, epinephrine, hydrochlorothiazide, and levothyroxine

6. This drug reduces the growth of facial hair in idiopathic hirsutism or hirsutism secondary to androgen excess

Questions 7-9: Match each structure with its embryonic origin.

    (A) Branchial groove 1

    (B) Branchial arch 1

    (C) Pharyngeal pouch 1

    (D) Branchial groove 2

    (E) Branchial arch 2

    (F) Pharyngeal pouch 2

    (G) Branchial groove 3

    (H) Branchial arch 3

    (I) Pharyngeal pouch 3

    (J) Branchial groove 4

    (K) Branchial arch 4

    (L) Pharyngeal pouch 4

7. Auditory tube

8. External acoustic meatus

9. Glossopharyngeal nerve

Questions 10-12: For each example of an unconscious attitude and the conscious attitude that "conceals" it, choose the defense mechanism to which it most closely corresponds.

    (A) Absolution

    (B) Companionship

    (C) Denial

    (D) Hate

    (E) Isolation

    (F) Jealousy

    (G) Love

    (H) Projection

    (I) Reaction formation

    (J) Undoing

10. I hate him (unconscious)–I love him (conscious)

11. I hate him (unconscious)–he hates me (conscious)

12. I hate him (unconscious)–I don't hate him (conscious)

Questions 13-14: For each description below, choose the type of leukemia with which it is most likely to be associated.

    (A) Acute lymphoblastic leukemia

    (B) Acute myeloblastic leukemia (M1)

    (C) Acute promyelocytic leukemia (M3)

    (D) Chronic lymphocytic leukemia

    (E) Hairy cell leukemia

13. The enzyme TdT is often present in leukemic cells; lymphadenopathy is characteristic and striking

14. It constitutes about 20 percent of childhood leukemias and relapse is common following chemotherapy in older adults

Questions 15-16: It is customary today to classify antiarrhythmic drugs according to their mechanism of action. This is best defined by intracellular recordings that yield monophasic action potentials. In the accompanying figure, the monophasic action potentials of (A) slow response fiber (SA node) and (B) fast Purkinje fiber are shown. For each description that follows, choose the appropriate drug with which the change in character of the monophasic action potential is likely to be associated.

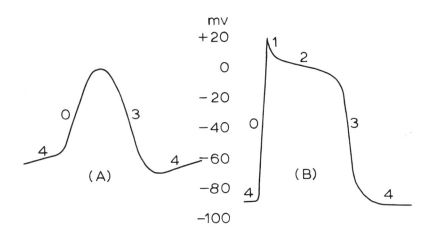

(A) Amiodarone (Cordarone)

(B) Digitalis

(C) Disopyramide (Norpace)

(D) Flecainide (Tambocor)

(E) Mexiletine (Mexitil)

(F) Nifedipine (Procardia)

(G) Propranolol (Inderal)

(H) Verapamil (Calan, Isoptin)

15. Moderate phase 0 depression and slow conduction; prolonged repolarization

16. Affects mainly phase 3; prolongs repolarization

Questions 17-18: Match each of the numbered descriptions of a hormone below with the appropriate lettered name of that hormone.

(A) Cortisol

(B) 1,25-Dihydroxycholecalciferol

(C) Epinephrine

(D) Estradiol

(E) Glucagon

(F) Insulin

(G) Luteinizing hormone

(H) Oxytocin

(I) Pancreatic polypeptide

(J) Progesterone

(K) Prolactin

(L) Secretin

(M) Somatostatin

(N) Thyrotropin-releasing hormone

(O) Thyroxine

(P) Vasopressin

17. Exerts its actions by binding to a plasma membrane receptor but does not seem to utilize a cyclic nucleotide second messenger mechanism

18. A water soluble hormone secreted by pancreatic A (α) cells

Match each numbered cytogenetic notation with the appropriate lettered phenotype.

    (A) Down syndrome

    (B) Down syndrome with possible atypical phenotype

    (C) Patau syndrome (the phenotype produced by an extra copy of chromosome 13)

    (D) Translocation carrier with normal phenotype

    (E) Turner syndrome

19. 47,XX,+der(2)(2pter→ 2q11::21p11→21qter)

20. 46,X,i(Xq)

Questions 21-22: Match the following descriptive phrases with the appropriate lettered type of cardiomyopathy.

    (A) Constrictive (restrictive) cardiomyopathy

    (B) Dilated (congestive) cardiomyopathy

    (C) Endomyocardial fibrosis

    (D) Hypertrophic cardiomyopathy

    (E) Secondary cardiomyopathy

21. Thirty percent of patients are prone to sudden cardiac death

22. May develop in patients with multiple myeloma

Questions 23-28: Match each description with the correct cell type.

    (A) Basophil

    (B) Connective tissue mast cell

    (C) Eosinophil

    (D) Fibroblast

    (E) Lymphocyte

    (F) Macrophage

    (G) Mucosal mast cell

    (H) Neutrophil

    (I) Plasma cell

23. Active in phagocytosis of antigen-antibody complex; otherwise not considered a major phagocyte

24. Found in submucosa of gastrointestinal tract; secretion of histamine, heparin, leukotrienes, and chemoattractants

25. Metachromatic staining cells near luminal surface of small intestinal epithelium; contains chondroitin sulfate rather than heparin; T-cell-dependent

26. Accumulation of these dead cells forms pus

27. Phagocytosis mediated by complement or IgG

28. Regulation of allergic reactions by negative feedback

Questions 29-30: Match each vitamin with the appropriate description.

(A) Excess amounts should be avoided when the patient is on levodopa

(B) Overdosage may lead to a psychotic state

(C) Improvement of vision especially in daylight might be attributable to this vitamin

(D) This vitamin is usually not included in the popular "one-a-day" vitamin preparations

(E) Retinoic acid is the natural form

(F) Acute intoxication with this vitamin causes hypertension, nausea and vomiting, and signs of increased CSF pressure

(G) This vitamin has hormonal functions

(H) This fat-soluble vitamin has mainly antioxidant properties

(I) In its water-soluble form, this fat-soluble vitamin is capable of producing kernicterus

29. Phytonadione

30. Calcitriol

Questions 31-33: For each numbered patient, select the lettered drug or agent most likely to cause the toxic effect described.

(A) Aluminum

(B) Bismuth

(C) Carbon monoxide

(D) Dapsone

(E) Ethylene glycol

(F) Gentamicin

(G) Lead

(H) Methanol

(I) Metronidazole

(J) Nalidixic acid

(K) Primaquine

(L) Sulfamethoxazole

(M) Sulfasalazine

(N) Tetracycline

31. A 53-year-old woman presents at your office with an *E. coli* urinary tract infection. During the course of treatment a blood sample is drawn from your patient and you discover she is experiencing hemolysis

32. A 3-year-old boy is brought to you for examination after he consumed an unknown liquid from a container he found in his family's garage. He exhibits central nervous system depression, acidosis, and suppressed respiration. On microscopic examination, you observe oxalate crystals in his urine. Besides supportive and corrective measures, you administer ethanol to the child

33. A distraught mother brings her 4-year-old daughter to you because the child is constipated, seems to have colic, and, most disturbing, has been exhibiting tremors. Upon examination, you also observe ataxia, weakness of extensor muscles, increased δ-aminolevulinic aciduria, and basophilic, stippled erythroblasts

Questions 34-35: For each disease, choose the sign with which it is most likely to be associated.

(A) Cowdry A intranuclear inclusions

(B) Hepatolenticular degeneration

(C) Lewy bodies

(D) Neurofibrillary tangles

(E) Optic nerve demyelination

(F) Verocay bodies

34. Idiopathic parkinsonism

35. Herpes simplex encephalitis

Questions 36-37: For each of the drugs below, select the most suitable description.

(A) Parenteral penicillin that is resistant to β-lactamase

(B) Oral penicillin that is resistant to β-lactamase

(C) Referred to as an extended-spectrum penicillin

(D) Chemically, the compound is a cephalosporin

(E) Related to ampicillin but with better oral absorption

(F) Administered intramuscularly and yields prolonged drug levels

(G) Cause of a disulfiram-like reaction

(H) Given parenterally and may cause elevation of serum sodium

(I) Cause of hypothrombinemia

36. Amoxicillin

37. Carbenicillin

Questions 38-42: The pedigree shown in the figure below contains individuals with Charcot-Marie-Tooth (CMT) disease, a neurologic disorder that produces dysfunction of the distal extremities with characteristic footdrop. Match the individuals in the pedigree with their probability of having an affected child with CMT.

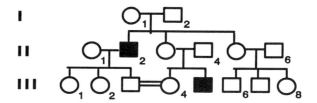

(A) 1 in 2

(B) 1 in 4

(C) 1 in 8

(D) 1 in 16

(E) Virtually 0

38. Individual II-2

39. Individual II-3

40. Individual II-5

41. Individual III-4

42. Individual III-8

7

Questions 43-45: Choose the lettered response that best matches each numbered bacterium.

(A) Secretes erythrogenic toxin that causes the characteristic signs of scarlet fever

(B) Produces toxin that blocks protein synthesis in an infected cell and carries a lytic bacteriophage that produces the genetic information for toxin production

(C) Produces at least one protein toxin consisting of two subunits, A and B, that cause severe spasmodic cough usually in children

(D) Requires cysteine for growth

(E) Secretes exotoxin that has been called "verotoxin" and "Shiga-like toxin"; infection is mediated by specific attachment to mucosal membranes

(F) Possesses *N*-acetylneuraminic acid capsule and adheres to specific tissues by pili found on the bacterial cell surface

(G) Has capsule of polyglutamic acid, which is toxic when injected into rabbits

(H) Synthesizes protein toxin as a result of colonization of vaginal tampons

(I) Causes spontaneous abortion and has tropism for placental tissue due to the presence of erythritol in allantoic and amniotic fluid

(J) Secretes two toxins, A and B, in large bowel during antibiotic therapy

(K) Has 82 polysaccharide capsular types; capsule is antiphagocytic; type 3 capsule (β-D-glucuronic acid polymer) most commonly seen in infected adults

43. *Streptococcus pyogenes*

45. *Brucella*

44. *Neisseria meningitidis*

Questions 46-47: Match the characteristic features with the appropriate acanthotic skin disease.

(A) Acanthosis nigricans

(B) Lichen planus

(C) Pemphigus vulgaris

(D) Psoriasis

(E) Verruca vulgaris

46. Parakeratosis and elongation of clubbed rete ridges and dermal papillae

47. Microabscesses in the stratum corneum

Questions 48-49: Match each numbered clinical situation below with the appropriate lettered risk figure.

    (A) 1/10,000

    (B) 1/600

    (C) 1/100

    (D) 1/10

    (E) 1

48. The risk for a newborn to have Down syndrome

49. The risk for a balanced translocation carrier to have a child with unbalanced chromosomes

Questions 50-52: For each of the following numbered agents, select the letter of its appropriate site of action in the acetylcholine system shown below.

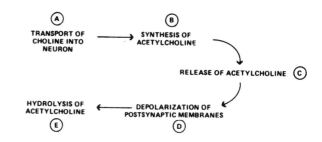

50. Botulinus toxin

51. Isofluorophate

52. Tubocurarine

Questions 53-54: The antigens and antibody in the questions below are associated with hepatitis. For each, choose the lettered description with which it is most likely to be associated.

    (A) Is usually the first viral marker detected in blood after HBV infection

    (B) May be the only detectable serologic marker during the early convalescent phase of an HBV infection ("window phase")

    (C) Appears in the blood soon after infection, rises to very high concentrations, and falls rapidly with the onset of hepatic disease

    (D) Found within the nuclei of infected hepatocytes and not generally in the peripheral circulation

    (E) Closely associated with hepatitis B infectivity and DNA polymerase activity

53. HBeAg

54. HBsAg

Questions 55-56: Many families of drugs consist of members that vary only with respect to substituents on a common ring structure. For each numbered type of pharmacologic effect that follows, select the lettered ring structure with which it is most likely to be associated.

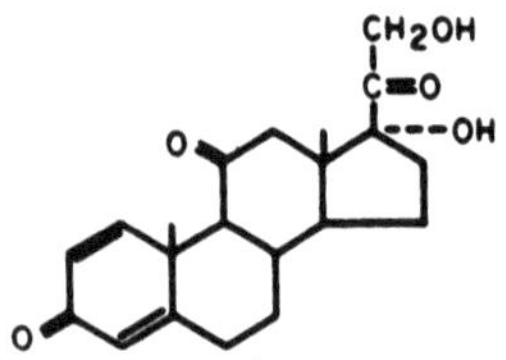

A

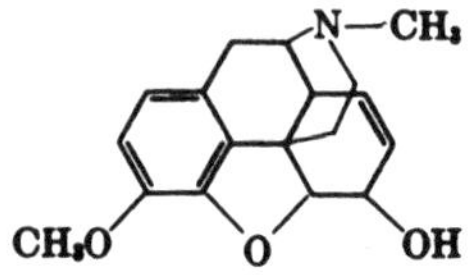

B

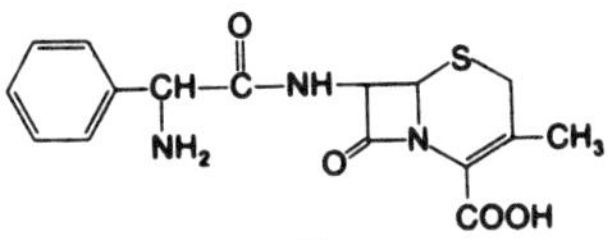

C

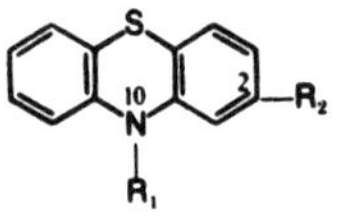

D

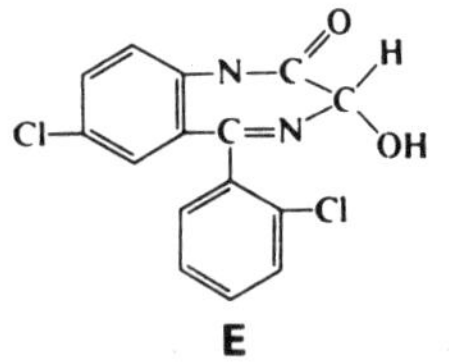

E

55. Antimicrobial

56. Corticosteroid antiinflammatory

Questions 57-58: For each outline of an experiment, select the form of statistical analysis that is most appropriate.

(A) Analysis of variance

(B) Chi-square

(C) Linear regression

(D) One-tail t-test

(E) Two-tail t-test

57. Does the inhibition of dopamine-sensitive adenylate cyclase have predictive value in assessing the potency of antipsychotic drugs? The data for each of 20 currently available antipsychotic drugs include an inhibition constant (K_i) that represents the drug concentration required to produce 50 percent inhibition of dopamine-stimulated cyclic AMP formation and an average clinical dose as a measure of potency

Drug	K_i (nM)	Approximate Equivalent Daily Dose (mg)
Chlorpromazine	50.0	100
Thioridazine	137.0	93
Fluphenazine	4.3	11
Trifluoperazine	18.0	6
•	•	•
•	•	•
•	•	•

58. Are black people who undergo a common surgical procedure more likely to be treated by a less experienced surgeon than are white people? The data represent hospital records from 340 patients undergoing gallbladder or hernia-repair operations in 10 randomly selected hospitals and include the patients' race (only black and white patients have been retained in the sample) and the status—resident or staff—of the surgeon

| Race of Patient | Status of Surgeon | |
	Staff	Resident
Black	31	63
White	31	215

59. The principal long-term storage form for energy in humans is

    (A) amino acids

    (B) cholesterol esters

    (C) fatty acids

    (D) glycogen

    (E) triacylglycerols

60. A 37-year-old woman with chills, fever, and headache is thought to have "atypical" pneumonia. Her history reveals that she raises chickens and that approximately two weeks ago lost a large number of them to an undiagnosed disease. Which of the following is the most likely diagnosis?

    (A) Anthrax

    (B) Leptospirosis

    (C) Psittacosis

    (D) Q fever

    (E) Relapsing fever

61. A 19-year-old woman is seen with a form of adenosis consisting of glands with clear cytoplasm that resembles that of the endocervix. This has been termed vaginal adenosis and characteristically precedes the development of which of the following?

    (A) Cervical carcinoma

    (B) Condyloma acuminatum

    (C) Clear cell carcinoma

    (D) Endometrial carcinoma

    (E) Squamous carcinoma of the vagina

62. A 48-year-old patient who has been suffering from recurrent headaches presents at your office. She tells you she has been applying pressure on the arteries of her neck to relieve the headache pain. Such occlusion of both carotid arteries between the heart and the carotid sinuses would be expected to produce

    (A) decreased activity of the vasomotor center

    (B) decreased central venous pressure

    (C) decreased heart rate

    (D) increased femoral arterial blood pressure

    (E) peripheral vasodilatation

63. Stroke volume is the volume of blood each ventricle of the heart ejects per beat. It will be increased by which of the following?

    (A) Increased systemic arterial blood pressure

    (B) Increased cardiac contractility

    (C) Increased heart rate

    (D) Reduced end diastolic volume

    (E) Decreased sympathetic stimulation

64. On the assumption that passive transport of the nonionized forms of the following drugs determines the rate of their absorption, which of the following drugs will be best absorbed in the small intestine?

    (A) Acetylsalicylic acid ($pK_a = 3.0$)

    (B) Ethacrynic acid ($pK_a = 3.5$)

    (C) Secobarbital ($pK_a = 7.8$)

    (D) Sulfamethoxazole ($pK_a = 5.6$)

    (E) Theophylline ($pK_a = 8.8$)

65. The figure below shows the change in mean blood pressure as a result of increasing doses of norepinephrine and the antagonism of this response by drugs X and Y. Based on the information provided in the diagram, which of the following statements is correct?

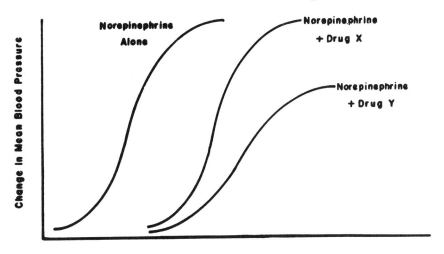

(A) Drug X is a more effective antagonist than drug Y

(B) Drug X is a more potent antagonist than drug Y

(C) Drug X shows the characteristics of competitive antagonism

(D) Drug Y shows the characteristics of competitive antagonism

(E) Drug Y is a more potent antagonist than drug X

---

66. A 23-year-old woman, when given a prescription for an oral contraceptive containing both progestational and estrogenic compounds, asks her physician about side effects. The adverse reactions of norethindrone include which of the following?

(A) Choriocarcinoma

(B) Hirsutism

(C) Multiple birth

(D) Venous thromboembolic disease

(E) Weight loss

67. Hormone-sensitive lipase is a cytoplasmic enzyme in adipocytes. Which of the following statements regarding activation of hormone-sensitive lipase in adipocytes is correct?

(A) It results in accumulation of monoglycerides and diglycerides in adipocytes

(B) It causes increased hydrolysis of cholesterol esters

(C) It is prevented by cortisol

(D) It is mediated by a cyclic AMP-dependent protein kinase

(E) It is stimulated by insulin

68. In the graph below of tubular loss of glucose versus tubular load of glucose the transport maximum (T_m) for glucose is located at point

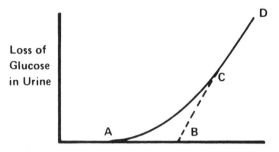

Tubular Load of Glucose

    (A) A

    (B) B

    (C) C

    (D) D

    (E) none of the above

69. In the typical life cycle of a trematode (e.g., *Schistosoma*), which of the following developmental forms enters the intermediate snail host?

    (A) Cercaria

    (B) Metacercaria

    (C) Micacidium

    (D) Redia

    (E) Schizont

70. Conduction velocity in myocardial cells is increased by

    (A) decreased membrane capacitance

    (B) decreased space constant

    (C) partially depolarized membrane

    (D) reduced fiber diameter

    (E) reduced slope of phase 0

71. The anterior hypothalamus contains neurons that monitor internal (local brain) temperature and activate appropriate thermoregulatory mechanisms when this temperature deviates from "normal." If, in an experimental animal, this area of the anterior hypothalamus is heated slightly, you would expect to see

    (A) cutaneous vasoconstriction

    (B) huddling or curling of the body

    (C) panting

    (D) piloerection

    (E) vigorous shivering

72. Isoenzymes from different tissues catalyze the same reaction. It is likely that the two reactions will share which of the following properties?

    (A) Activators and inhibitors

    (B) Cofactors

    (C) Equilibrium constants

    (D) Optimum pH

    (E) V_{max} values

73. A 27-year-old woman is being treated for infertility apparently due to a lack of ovulation. Which of the following substances enhances the probability of ovulation by blocking the inhibitory effect of estrogens and thus stimulating the release of gonadotropin from the pituitary?

    (A) Clomiphene

    (B) Diethylstilbestrol

    (C) Ethinyl estradiol

    (D) Oxymetholone

    (E) Progesterone

74. A 20-year-old woman is being evaluated because she has never menstruated. She is of average intelligence and short stature. You suspect Turner's syndrome. A buccal smear shows most cells having no Barr bodies, but some cells having one Barr body. Which of the following best explains this finding?

(A) Laboratory error, patient is a female (XX)

(B) The patient is a male (XY)

(C) Classic Turner's syndrome (XO)

(D) Turner mosaic

(E) Klinefelter's syndrome

75. Which of the curves shown below depicts the blood levels of a drug administered by rapid intravenous injection?

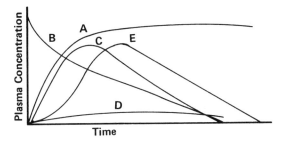

(A) Curve A

(B) Curve B

(C) Curve C

(D) Curve D

(E) Curve E

76. A 46-year-old woman has experienced severe abdominal pain, nausea, and vomiting for two days. The pain, which is sharp and constant, began in the epigastric region and radiated bilaterally around the chest to just below the scapulae. Subsequently, the pain became localized in the right hypochondrium. Palpation reveals marked tenderness in the right hypochondriac region and some rigidity of the abdominal musculature. An x-ray shows numerous calcified stones in the region of the gallbladder. Diffuse pain referred to the epigastric region and radiating circumferentially around the chest is the result of afferent fibers that travel via which of the following nerves?

(A) Greater splanchnic

(B) Intercostal

(C) Phrenic

(D) Vagus

(E) None of the above

77. An 18-month-old boy is diagnosed as having bacterial meningitis. What is the most common cause of this condition?

(A) *Escherichia coli* type K

(B) *Haemophilus influenzae*

(C) *Klebsiella pneumoniae*

(D) *Neisseria meningitidis*

(E) *Streptococcus pyogenes*

78. Mycoplasmas are bacterial cells that

(A) are resistant to penicillin

(B) are resistant to tetracycline

(C) fail to reproduce on artificial media

(D) have a rigid cell wall

(E) stain well with Gram stain

79. During a physical examination, an 18-year-old female is found to have splenomegaly. Serologic examination reveals an elevated white blood cell count (including atypical lymphocytes) and heterophil antibodies. She probably has

    (A) infectious mononucleosis

    (B) lymphocytic choriomeningitis

    (C) mumps

    (D) parainfluenza

    (E) rubella

80. An increased affinity of hemoglobin for O_2 may result from which of the following changes within erythrocytes?

    (A) decreased H^+ concentration

    (B) elevated CO_2 levels

    (C) elevated temperature

    (D) increased 2,3-diphosphoglycerate (DPG) levels

    (E) increased H^+ concentration

81. Causes of interstitial lung disease include occupational exposure to inorganic dust (asbestos, silica), gases, aerosols, and organic dust, in addition to drugs (bleomycin, busulfan, nitrofurantoin) and infections (cytomegalovirus and tuberculosis). The major interstitial lung diseases of unknown cause include sarcoidosis and idiopathic pulmonary fibrosis (Hamman-Rich syndrome). Irrespective of the cause of interstitial lung disease, the initial pathologic event is generally accepted to be

    (A) bronchopneumonia

    (B) bronchitis

    (C) alveolitis

    (D) bronchiolitis

    (E) pleuritis

82. In urea synthesis, which of the following reactions requires adenosine triphosphate (ATP)?

    (A) Arginine → ornithine + urea

    (B) Citrulline + aspartate → argininosuccinate

    (C) Fumarate → malate

    (D) Oxaloacetate + glutamate → aspartate + α-ketoglutarate

    (E) None of the above

83. The structure shown below is a member of which of the following drug groups?

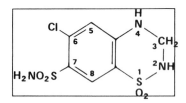

    (A) Carbonic anhydrase inhibitors

    (B) Osmotic diuretics

    (C) Potassium-sparing diuretics

    (D) Thiazide diuretics

    (E) Xanthine diuretics

84. A 14-year-old boy was taken to the doctor with a fever of unknown origin. History revealed that his family raises goats and he has been helping to care for them for the past year. His fever would most likely be caused by which of the following organisms?

    (A) *Brucella melitensis*

    (B) *Francisella tularensis*

    (C) *Leptospira interrogans*

    (D) *Pseudomonas pseudomallei*

    (E) Vesicular stomatitis virus

85. The point called T_m *(melting temperature)* for double-stranded DNA is represented by which of the letters in the graph below?

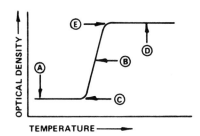

(A) A

(B) B

(C) C

(D) D

(E) E

86. Which of the following vitamins is most likely to become deficient in a person who is completely carnivorous?

(A) Ascorbic acid

(B) Cobalamin

(C) Niacin

(D) Pantothenic acid

(E) Thiamine

87. Which of the following viruses is the leading cause of the croup syndrome in young children and, when infecting mammalian cells in culture, will heme-absorb red blood cells?

(A) Adenovirus

(B) Group B coxsackievirus

(C) Parainfluenza virus

(D) Rhinovirus

(E) Rotavirus

88. Excessive drinking of alcohol by college students is a serious social problem in the United States. College students who are heavy alcohol drinkers

(A) are more likely to major in natural science than the social sciences

(B) tend to be deeply religious

(C) tend to come from large metropolitan rather than rural areas

(D) use nonalcoholic drugs less frequently than do nondrinkers

(E) tend to come from the southwestern United States

89. A 21-year-old man is found to have Hodgkin's disease, clinical stage IIA, with lymphocyte predominance. A true statement concerning his case is which of the following?

(A) He will have fever, pruritus, and weight loss

(B) Numerous Reed-Sternberg cells and eosinophils will be found on biopsy

(C) The prognosis is poor

(D) His involved lymph nodes will be on the same side of the diaphragm

(E) Reed-Sternberg cells will be absent in the biopsy

90. About 30 to 35 percent of the people who cannot sleep have a relatively simple organic cause for the problem. One of the most common organic explanations for a sleep disturbance in an otherwise healthy person is

(A) suppressed REM sleep

(B) suppressed stage 4 sleep

(C) the accumulation of hepatic enzymes

(D) the consequences of aging

91. The juxtamedullary nephrons of the kidney have relatively long loops of Henle and constitute about 15 percent of the total nephrons in the human. The fluid in a juxtamedullary nephron

(A) enters the collecting duct hypertonic to plasma

(B) enters the collecting duct isotonic to plasma

(C) enters the loop of Henle hypertonic to plasma

(D) leaves the loop of Henle hypertonic to plasma

(E) leaves the loop of Henle isotonic to plasma

92. A cardiac patient treated with digoxin is also given quinidine. When quinidine and digoxin are administered concurrently, which of the following effects does quinidine have on digoxin?

(A) It antagonizes the effect of digoxin on the AV node

(B) It decreases the absorption of digoxin from the GI tract

(C) It increases the plasma concentration of digoxin

(D) It prevents the metabolism of digoxin

(E) It reduces inhibition of Na^+-K^+-stimulated ATPase by digoxin

93. The amount of tension that a whole muscle can produce is greatest in which of the following situations?

(A) Maximal complete tetanus

(B) Maximal incomplete tetanus

(C) Same in all situations (all or none)

(D) Single twitch response

(E) Submaximal complete tetanus

94. A 52-year-old automobile salesman is being treated for a subdural hematoma after coming to the hospital emergency room complaining of an overwhelming headache. When blood enters the potential space between the arachnoid and dura, a subdural hematoma forms. This occurs most frequently in the

(A) supracerebellar region

(B) infracerebellar region

(C) cerebellopontine angle

(D) pituitary region

(E) cerebral hemisphere convexities

95. Dipicolinic acid, which has the structure shown below, is a key component of

(A) bacterial flagella

(B) bacterial pili

(C) bacterial spores

(D) eukaryotic cilia

(E) eukaryotic flagella

96. A disorder resulting from a single gene defect that may produce severe mental problems is

(A) manic-depressive psychosis

(B) dyslexia

(C) phenylketonuria

(D) Porter's syndrome

(E) Down's syndrome

97. A 63-year-old woman has breast cancer with metastases to soft tissues. The disseminated breast cancer is found to be estrogen-receptor positive. These tumors are most likely to respond to the antiestrogenic effect of

(A) bleomycin

(B) dacarbazine

(C) diethylstilbestrol

(D) fluoxymesterone

(E) tamoxifen

98. The nephrotic syndrome is strongly associated with many varieties of glomerulonephritis. Which of the following, however, does not usually lead to the nephrotic syndrome?

(A) Acute tubular necrosis

(B) Amyloidosis

(C) Lipoid nephrosis

(D) Membranoproliferative glomerulonephritis

(E) Membranous glomerulonephritis

99. Disseminated candidiasis can be life-threatening either as a primary infection in immunosuppressed patients or as a secondary infection of the lungs, kidneys, and other organs in persons who have tuberculosis or cancer. Although candidiasis of the oral cavity (thrush) usually is controlled by the administration of nystatin, the disseminated or systemic form of candidiasis requires vigorous therapy with

(A) amphotericin B

(B) chloramphenicol

(C) interferon

(D) penicillin

(E) thiabendazole

100. A 52-year-old male patient complained to his dentist about a draining lesion in his mouth. A Gram stain of the pus showed a few gram-positive cocci, leukocytes, and many branched, gram-positive rods. The most likely cause of the disease is

(A) *Actinomyces israelii*

(B) *Corynebacterium diphtheriae*

(C) *Propionibacterium acnes*

(D) *Staphylococcus aureus*

(E) *Streptococcus mutans*

101. When a young, healthy person gets out of bed in the morning and stands, the venous pressure in her dural sinuses normally falls within which of the following ranges?

(A) Subatmospheric to 0 mmHg

(B) 0 to 5 mmHg

(C) 5 to 10 mmHg

(D) 10 to 20 mmHg

(E) 20 to greater than 20 mmHg

102. The Cori cycle is best described as

(A) the transfer of reducing equivalents from the cytosol to the mitochondrial matrix

(B) the movement of ammonia in the form of glutamine from muscle to liver causing the formation of urea

(C) the transfer of lactate from muscle to the liver resulting in the formation of glucose

(D) the transfer of alanine formed from pyruvate in muscle to the liver where it is converted to glucose

(E) the formation of glycogen from glucose 1-phosphate and the phosphorylation of glycogen back to glucose 1-phosphate

103. A linear pattern of immunoglobulin deposition along the glomerular basement membrane that can be demonstrated by immunofluorescence is typical of

(A) diabetic glomerulopathy

(B) Goldblatt's kidney

(C) Goodpasture's syndrome

(D) lupus nephritis

(E) renal vein thrombosis

104. The methyl donor in most biosynthetic processes is

(A) homocysteine

(B) N^5-methyltetrahydrofolate

(C) methionine

(D) S-adenosylmethionine

(E) tetrahydrofolate

105. Observation of a histologic preparation of muscle indicates the presence of neuromuscular junctions, cross-striations, and peripherally located nuclei. The use of histochemistry demonstrates a strong staining reaction for succinic dehydrogenase. The same tissue prepared for electron microscopy demonstrates many mitochondria in rows between myofibrils and underneath the sarcolemma. The best description of this tissue is

(A) cardiac muscle

(B) fibers that contract rapidly, but are incapable of sustaining continuous heavy work

(C) red muscle fibers

(D) smooth muscle

(E) white muscle fibers

106. The histological findings shown below of nipple skin are most commonly found in association with

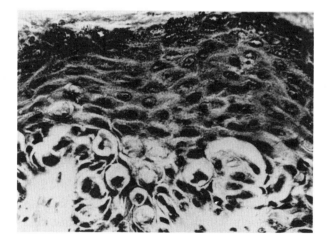

(A) a systemic vasculitic reaction

(B) no lesion, this is normal skin

(C) fibrocystic disease

(D) an underlying carcinoma

(E) a tender breast in a lactating woman

20

107. A man who has a deep laceration of the scalp with profuse bleeding is seen in the emergency room. His epicranial aponeurosis (galea aponeurotica) is penetrated, resulting in a severely gaping wound. The structure seen underlying the epicranial aponeurosis is

    (A) a layer containing blood vessels

    (B) the frontal bone

    (C) the dura mater

    (D) the periosteum (pericranium)

    (E) the tendon of the epicranial muscles (occipitofrontalis)

108. During the course of a routine physical examination of a 42-year-old woman, you order a blood workup including blood gases. She has a normal hematocrit and hemoglobin concentration and normal venous and arterial P_{O_2}. The hemoglobin-oxygen saturation of blood entering her right ventricle is approximately

    (A) 55 percent

    (B) 65 percent

    (C) 75 percent

    (D) 85 percent

    (E) 95 percent

109. The presence of cytoplasmic inclusion bodies in nerve cells of the spinal cord and brain (Negri bodies) is characteristic of

    (A) aseptic meningitis

    (B) congenital rubella

    (C) infectious mononucleosis

    (D) mumps

    (E) rabies

110. Reverse transcriptase is an enzyme used by RNA-containing viruses such as HIV. It catalyzes the polymerization of

    (D) ribonucleoside triphosphates using a DNA template

    (B) deoxyribonucleoside triphosphates using an RNA template

    (E) ribonucleoside triphosphates using an RNA template

    (A) deoxyribonucleoside triphosphates independently of a template

    (C) ribonucleoside and deoxyribo-nucleoside triphosphates to form mixed polymers

111. You are about to prescribe an antibiotic drug to a patient whom you know is a regular, heavy user of an antacid compound containing both aluminum and calcium salts. These divalent cations might inhibit the intestinal absorption of which of the following agents?

    (A) Chloramphenicol

    (B) Erythromycin

    (C) Isoniazid

    (D) Phenoxymethyl penicillin

    (E) Tetracycline

112. The Filoviridae are a family of negative-sense, single-stranded RNA viruses. Which of the following viruses belongs to this family and causes hemorrhagic fever?

    (A) Dengue virus

    (B) Ebola virus

    (C) Lassa fever virus

    (D) Parvovirus

    (E) Yellow fever virus

113. A 58-year-old man is undergoing treatment for renal insufficiency. Recently, he has developed very high plasma concentrations of urea (uremia). Which one of the following is the most likely reason he became uremic?

   (A) Glomerular filtration rate is decreased

   (B) Hepatic synthesis of urea is increased

   (C) Reabsorption of urea in the proximal tubules is decreased

   (D) Secretion of urea by distal tubules is decreased

   (E) Synthesis of urea in the tubules is increased

114. Glycerol available for participation in fatty acid esterification in adipocytes

   (A) essentially is derived from extracellular sites and transported to adipocytes

   (B) for the most part is derived from glucose

   (C) is formed by gluconeogenesis

   (D) is produced at a lower rate in the presence of insulin

   (E) primarily is obtained from phosphorylation of glycerol by glycerol kinase

115. In humans, the excreted end product of purine metabolism is

   (A) allantoic acid

   (B) orotic acid

   (C) urea

   (D) uric acid

   (E) xanthine

116. The rate of emptying of the stomach is regulated to prevent movement of gastric contents into the duodenum at a rate faster than they can be processed by the small bowel. Which of the following would most *slow* the rate of gastric emptying?

   (A) Activity in vagal neurons

   (B) A decrease in duodenal pH

   (C) Distention of the stomach

   (D) Eating a carbohydrate-rich meal

   (E) Increased secretion of gastrin

117. Many disorders that present in adult life, such as coronary artery disease and hypertension, are multifactorial traits. A multifactorial trait results from

   (A) the interaction between the environment and a single gene

   (B) the interaction between the environment and multiple genes

   (C) multiple postnatal environmental factors

   (D) multiple pre- and postnatal environmental factors

   (E) multiple genes independent of environmental factors

118. Ascorbic acid is an essential vitamin because it is

   (A) necessary for CO_2 fixation

   (B) required for conversion of proline to hydroxyproline

   (C) a precursor for coenzyme A

   (D) a precursor for NADP

   (E) a precursor for flavin mononucleotide

119. Epstein-Barr virus is the etiologic agent of infectious mononucleosis. It has also been implicated in Burkitt's lymphoma and nasopharyngeal carcinoma. What other virus has been shown to cause infectious mono-nucleosis-like disease?

(A) Adenovirus

(B) Cytomegalovirus

(C) Herpes simplex type 1

(D) Respiratory syncytial virus

(E) Rubella virus

120. Presynaptic inhibition is an important regulatory mechanism at many neural syn-apses in the central nervous system. The crucial factor operating in the terminal under presynaptic inhibition is a

(A) blockage of the voltage-sensitive K^+ channels

(B) decrease in the Ca^{2+} flux

(C) decrease in Na^+ flux

(D) hyperpolarization of the terminal

(E) plastic change in the type of transmitter being released

121. A 37-year-old woman presents with a lump in the upper outer quadrant of the left breast, which shows a wide spectrum of benign breast disease on pathologic exami-nation. Which of the following is considered to indicate the greatest risk for subsequent carcinoma of the breast?

(A) Epithelial hyperplasia of the ducts

(B) Florid papillomatosis

(C) Intraductal papillomatosis

(D) Marked apocrine metaplasia

(E) Sclerosing adenosis

122. During early inflammation following margination of leukocytes, an important adhesive surface protein on leukocytes is

(A) complement factor 5a (C5a)

(B) endothelial leukocyte adhesion molecule (ELAM-1)

(C) interleukin 1 (IL-1)

(D) leukocyte function antigen 1 (LFA-1)

(E) tumor necrosis factor (TNF)

123. Intervertebral disks have a characteristic pattern of herniation into the intervertebral foramen because the

(A) annulus fibrosus of the intervertebral disk is attenuated on both sides in the posterolateral regions

(B) intervertebral disks are narrower anteriorly than posteriorly

(C) posterior longitudinal ligament is stronger posteriorly than posterolaterally

(D) vertebral bodies and intervertebral disks are reinforced posteriorly by the ligamentum flavum

(E) vertebral bodies and intervertebral disks are weakly reinforced ventrally by the anterior longitudinal ligament

124. Phosphorylation sites in the mitochondrial respiratory chain may occur between

(A) coenzyme Q and cytochrome b

(B) NAD and an iron-sulfur protein

(C) NAD and flavoprotein

(D) pyruvate and NAD

23

Questions 125-126.: Measurement of the closing volume is a sensitive test of airway disease. An 18-year-old man expires to residual volume and then inspires to total lung volume. At the beginning of this inspiration, a small quantity of helium is injected into the inspired gas. The patient then expires to residual volume and the curve appearing below is produced.

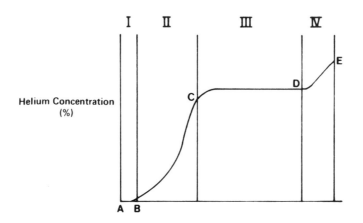

125. This patient's closing volume is measured at point

    (A) A

    (B) B

    (C) C

    (D) D

    (E) E

126. Normally, you would expect this patient's closing volume to be

    (A) 10 percent of his vital capacity and to decrease as he grows older

    (B) 10 percent of his vital capacity and to increase as he grows older

    (C) 40 percent of his vital capacity and to decrease as he grows older

    (D) 40 percent of his vital capacity and to increase as he grows older

    (E) 90 percent of his vital capacity and to decrease as he grows older

127. The major cause of death among adolescents and young adults is

    (A) homicide

    (B) illicit drug abuse

    (C) infections

    (D) motor vehicle accidents

    (E) suicide

128. A low level of the activity of which one of the following enzymes would be a sensitive indicator of thiamine deficiency?

    (A) 2,3-Diphosphoglycerate mutase

    (B) Glucose-6-phosphate dehydrogenase

    (C) Transaldolase

    (D) Transketolase

129. A premature infant girl was mistakenly treated with sulfonamides and developed severe jaundice, an absent startle reflex, abnormal eye movements, and opisthotonus. She also had an unusually high-pitched cry. Kernicterus, this particular form of jaundice, is due to which of the following effects of the sulfonamides?

(A) Deposition of crystalline aggregates in the kidney

(B) Displacement of bound bilirubin from albumin

(C) Enhanced synthesis of bilirubin

(D) Inhibition of bilirubin degradation

(E) Inhibition of urinary excretion of bilirubin

130. Which of the following viruses causes an acute febrile rash and produces disease in immunocompetent children, but has been associated with transient aplastic crisis in persons with sickle cell disease?

(A) Herpes simplex

(B) Parvovirus

(C) Rubella

(D) Rubeola

(E) Varicella zoster

131. Which of the following substrates can make a net contribution to blood glucose via gluconeogenesis?

(A) Leucine

(B) Pyruvate

(C) Palmitate

(D) Acetyl coenzyme A

132. A 27-year-old female patient presents with a request for testing for human immunodeficiency virus (HIV) because of a weekend fling of promiscuous sexual activity 2 weeks ago. Given that time frame of possible infection, what is the best, most cost-efficient test to order?

(A) Enzyme immunoassay for HIV antibody

(B) Enzyme immunoassay for HIV antigen

(C) HIV Western blot

(D) Polymerase chain reaction for HIV

(E) HIV culture

133. The movement of substances across a capillary wall can occur by diffusion, filtration, or intracellular vesicular transport. If the concentration of a substance within capillary blood decreases linearly along the length of a capillary, which of the following statements regarding its movement from capillary blood to interstitial fluid is correct?

(A) If the plasma oncotic pressure decreases, there will be decreased movement from capillary blood to interstitial fluid

(B) Movement from capillary blood to interstitial fluid will be increased if the substance's plasma concentration increases

(C) Rate of movement from capillary blood to interstitial fluid is inversely related to the molecular size of the substance

(D) There will be increased movement from capillary blood to interstitial fluid if capillary blood flow increases

(E) Under normal flow conditions, the substance will reach equilibrium between capillary blood and interstitial fluid

134. The term *amino acid activation* refers to

  (A) the formation of an ester linkage between the alpha amino group of an amino acid and the 2′ or 3′ hydroxyl group of a transfer RNA

  (B) the formation of an ester linkage between the alpha carboxyl group of an amino acid and the 2′ or 3′ hydroxyl group of a transfer RNA

  (C) the formation of the peptide bond in protein synthesis

  (D) the intermediate formation of a coenzyme A derivative of an amino acid, before its linkage to a transfer RNA

135. A very important mediator of the systemic effects of inflammation is

  (A) gamma interferon (γ-IFN)

  (B) beta tumor necrosis factor (β-TNF)

  (C) interleukin 1 (IL-1)

  (D) interleukin 2 (IL-2)

  (E) interleukin 3 (IL-3)

136. A patient has been on long-term, high-dosage antibiotic therapy. She begins to complain of vertigo, an inability to perceive termination of movement, and difficulty in sitting or standing without visual clues. She also relates that it is becoming difficult to hear certain sounds. These signs and symptoms are the result of toxic reactions of which of the following drugs?

  (A) Amphotericin B

  (B) Isoniazid

  (C) Penicillin

  (D) Streptomycin

  (E) Tetracycline

137. The cell component responsible for packaging, and perhaps producing, glycosylated cellular products is the

  (A) Golgi apparatus

  (B) mitochondrion

  (C) nucleolus

  (D) rough endoplasmic reticulum

  (E) smooth endoplasmic reticulum

138. Within a few hours after birth, a newborn's ductus arterious constricts. It finally closes completely within a few weeks. When it fails to close spontaneously, surgical ligation may be necessary. Which of the following statements regarding the ductus arteriosus is correct?

  (A) A physiologic left-to-right shunt in the fetal circulation is maintained by the open ductus arteriosus

  (B) Closure of the ductus arteriosus is stimulated by a rise in P_{O_2} in the blood flowing through it

  (C) It is not uncommon for the ductus to close 2 to 4 weeks before birth

  (D) Patency of the ductus in the fetus is inhibited by prostacyclin

  (E) The flow of blood from the placenta to the vena cava is reduced following ductus closure

139. Which of the following social variables is most closely linked to infant mortality?

  (A) Economic status of the parents

  (B) Education of the father

  (C) Education of the mother

  (D) Marital status of the parents

  (E) Occupation of the father

140. In the figure shown below, Curve 3 is the oxygen dissociation curve of hemoglobin under normal physiologic conditions. If the 2,3-diphosphoglycerate concentration were increased, which oxygen dissociation curve would be obtained for hemoglobin?

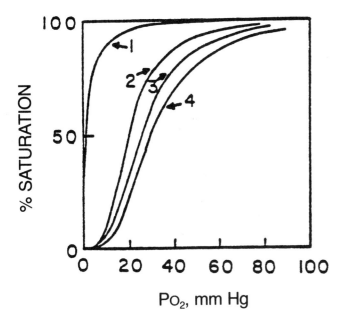

(A) Curve 1
(B) Curve 2
(C) Curve 3
(D) Curve 4

141. Malignant tumors of the ovary most commonly occur in women between the ages of 40 and 64. The majority of these malignant tumors of the ovary arise from

(A) hilar cells
(B) ovarian germ cells
(C) stromal cells
(D) surface epithelium
(E) urogenital stem cells

142. *E. coli* are grown in a medium containing equimolar concentrations of glucose and lactose. Which statement below best describes the cells' use of nutrients?

(A) Glucose is not used because a product of the lac operon blocks its entry into the cell
(B) Glucose is used preferentially because the concentration of cyclic AMP is low
(C) Lactose is used preferentially because the products of the lac operon are present
(D) Lactose is used preferentially because it contains two hexoses per molecule
(E) The two sugars are used about equally because both enter the cells with equal facility

143. A tube of monkey kidney cells is inoculated with nasopharyngeal secretions. During the next 7 days, no cytopathic effects (CPEs) are observed. On the eighth day, the tissue culture is infected accidentally with a picornavirus; nevertheless, the culture does not develop CPEs. The most likely explanation of this phenomenon is that

(A) monkey kidney cells are resistant to CPEs
(B) picornavirus does not produce CPEs
(C) picornavirus does not replicate in monkey kidney cells
(D) the nasopharyngeal secretions contained hemagglutinins
(E) the nasopharyngeal secretions contained rubella virus

Questions 144-145: During a regular, annual physical examination, a healthy 42-year-old airline pilot undergoes vigorous exercise on a treadmill.

144. During the treadmill exercise, you would expect the pulmonary blood flow to increase. Which of the following statements about that increase is correct?

(A) The percentage of increase in flow is greater in the bases of the lungs than in the apices

(B) The increase in flow is caused by a greater-than-fivefold increase in pulmonary arterial pressure

(C) The increase in pulmonary blood flow is less than the increase in systemic blood flow

(D) The increase in pulmonary blood flow is accommodated by dilation of pulmonary arterioles and capillaries

(E) The increase in pulmonary blood flow is caused by sympathetic nerve stimulation of the pulmonary vasculature

145. The blood flow in the skeletal muscles should increase tremendously during vigorous treadmill exercise. Which of the following factors contributes to the increased skeletal muscle blood flow?

(A) Activity of parasympathetic nerve fibers to the heart increases

(B) Circulating epinephrine constricts skeletal muscle blood vessels

(C) Local metabolites dilate skeletal muscle blood vessels

(D) Resistance of the veins and venules decreases

(E) Total peripheral resistance increases

146. A 5-year-old girl is brought to the emergency room with a rash and illness characteristic of chickenpox. One member of the medical team attending this patient is a pregnant nurse with no history of chickenpox. The decision is made to administer zoster immune globulin to the nurse, but the suggestion is made that a serum sample be drawn prior to this to determine her true immune status. The test necessary to determine immune status for varicella-zoster virus is

(A) complement fixation

(B) fluorescent antibody to membrane antigen (FAMA)

(C) direct fluorescent antibody

(D) indirect fluorescent antibody

(E) enzyme immunoassay

147. The hexose monophosphate shunt includes which one of the following enzymes?

(A) Alpha-glucosidase

(B) Fumarase

(C) Glucose-6-phosphate dehydrogenase

(D) Hexokinase

(E) Malic dehydrogenase

148. In general, the percentage of the cardiac output flowing to a particular organ is related to the metabolic activity of that organ in comparison with the other organs of the body. In which of the following organs is blood flow greater than might be predicted by its metabolic activity?

(A) Brain

(B) Heart

(C) Intestine

(D) Kidney

(E) Skeletal musculature

149. A 66-year-old man with no previous significant illness presents with back pain. The patient had felt well except for an increase in fatigue over the past few months. He suddenly felt severe low back pain while raising his garage door. Physical examination reveals a well-developed white male in acute pain. His pulse is 88 beats per minute and blood pressure is 150/90 mmHg. The conjunctivae are pale. There is marked tenderness to percussion over the lumbar spine. The following laboratory data are obtained: hemoglobin 11.0 g/dL (normal 13 to 16 g/dL), serum calcium 12.3 mg/dL (normal 8.5 to 11 mg/dL), abnormal serum protein electrophoresis with a monoclonal IgG spike, urine positive for Bence Jones protein, and abnormal plasma cells in bone marrow. X-rays reveal lytic lesions of the skull and pelvis and a compression fracture of the lumbar vertebrae. Your diagnosis would be

(A) hypoparathyroidism

(B) multiple myeloma

(C) osteomalacia

(D) osteoporosis

(E) Paget's disease

150. In a healthy 41-year-old man, oxygen consumption was measured at 700 mL/min. Pulmonary artery oxygen content was 140 mL per liter of blood and brachial artery oxygen content was 210 mL per liter of blood. Cardiac output was which of the following?

(A) 4.2 L/min

(B) 7.0 L/min

(C) 10.0 L/min

(D) 12.6 L/min

(E) 30.0 L/min

151. Haloperidol is a potent antipsychotic. The mechanism of action of haloperidol is the blockade of

(A) α_2-adrenergic receptors

(B) dopamine receptors

(C) γ-aminobutyric acid (GABA) receptors

(D) glutamate receptors

(E) serotonin

152. A 43-year-old patient reported to his physician that he was bitten by a tiny tick 3 to 4 weeks ago. He noted a confluent red rash around the tick bite followed by a flu-like illness. Which of the following laboratory tests would be likely, at this time, to confirm that the patient has Lyme disease?

(A) *Borrelia burgdorferi* specific IgG antibody (EIA)

(B) *Borrelia burgdorferi* specific IgM antibody (EIA)

(C) Cerebrospinal fluid (CSF) protein analysis

(D) Culture of blood for *Borrelia burgdorferi*

(E) Western blot analysis of specific IgG antibody response

153. A complete electrocardiogram evaluation is ordered for a 64-year-old man whom you suspect is suffering from heart disease. His electrocardiogram will not be capable of detecting abnormalities in

(A) atrioventricular conduction

(B) cardiac contractility

(C) cardiac rhythm

(D) coronary blood flow

(E) the position of the heart in the chest

154. Two allosterically regulated enzymes of the pathway from lactate to glucose are

- (A) phosphofructokinase and enolase
- (B) phosphofructokinase and lactate dehydrogenase
- (C) pyruvate carboxylase and fructose 1,6-bisphosphatase
- (D) pyruvate carboxylase and pyruvate dehydrogenase
- (E) triose phosphate isomerase and lactate dehydrogenase

155. If 91 percent of quinine, a weak base, is ionized in the blood, the pK_a of the drug is approximately

- (A) 2.6
- (B) 4.7
- (C) 6.4
- (D) 8.4
- (E) 9.3

156. A 24-year-old woman was hospitalized after an automobile accident. Her wounds became infected and she was treated with tobramycin, carbenicillin, and clindamycin. Five days after antibiotic therapy was initiated, she developed severe diarrhea and pseudomembranous enterocolitis. Antibiotic-associated diarrhea and the more serious pseudomembranous enterocolitis can be caused by

- (A) *Bacteroides fragilis*
- (B) *Clostridium difficile*
- (C) *Clostridium perfringens*
- (D) *Clostridium sordellii*
- (E) *Staphylococcus aureus*

157. Injury to the sciatic nerve by hypodermic injection can be avoided if the needle is inserted into the upper lateral quadrant of the buttock. This is because the sciatic nerve has which of the following anatomic relationships?

- (A) It passes lateral to the gluteus medius muscle
- (B) It emerges from the lesser sciatic foramen and extends inferiorly to the popliteal fossa
- (C) It passes into the thigh between the greater trochanter and ischial tuberosity
- (D) It travels down the medial aspect of the thigh to innervate the leg and foot
- (E) It typically passes superior and posterior to the piriformis muscle

158. A 72-year-old patient in the intensive care unit had multiple intravenous lines, including a Hickman catheter. The patient became febrile (40.6°C [105°F]) and multiple blood cultures were drawn. Two days later subculture of three blood-culture bottles revealed "diphtheroids." An antibiotic susceptibility test was performed by mistake on these alleged contaminants and revealed that the isolates were only susceptible to vancomycin. The report from the laboratory should read

- (A) "*Corynebacterium diphtheriae* isolated"
- (B) "diphtheroids—suspect skin contamination"
- (C) "no significant growth"
- (D) "suggest repeat specimen"
- (E) "unconfirmed *Corynebacterium JK (C. jaekium)* isolated"

159. The ovarian lesion shown in the photomicrograph below is

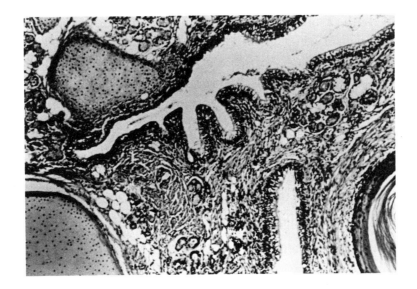

(A) chronic salpingitis

(B) a cystic teratoma

(C) an ectopic pregnancy

(D) a granulosa cell tumor

(E) metastatic squamous cell carcinoma

---

160. In appropriately stained interphase cells, a mass of heterochromatin may be seen lying against the nuclear membrane. This heterochromatin mass in nucleated cells is known as the Barr body. Which of the following statements is true of the Barr body?

(A) It represents the inactivated Y chromosome of males that remains condensed during interphase

(B) Its absence in a buccal smear chromatin test indicates definitively that the patient is female

(C) It would have an identical appearance in a buccal smear chromatin test in a patient with Turner's syndrome and in a normal female

(D) It is found exclusively in the germ cells of the gonads

(E) It may be used for determination of chromosomal sex and abnormalities of X-chromosome number

161. People with genetically low levels of *N*-acetyltransferase are more prone than normal persons to develop a lupus erythematosus-like syndrome following administration of which of the following drugs?

(A) Digitoxin

(B) Procainamide

(C) Propranolol

(D) Reserpine

(E) Selegiline

162. Aspirin inhibits which of the following enzymes?

(A) Cyclooxygenase

(B) Lipoprotein lipase

(C) Lipoxygenase

(D) Phospholipase A_2

(E) Phospholipase D

163. A medical laboratory technician recovering from hepatitis B develops hematuria, proteinuria, and red cell casts in the urine. Examination of this patient's kidneys would be most likely to reveal

   (A) amyloid deposits in the glomerular basement membrane

   (B) granular deposits of antibodies in the glomerular basement membrane

   (C) IgG linear fluorescence along the glomerular basement membrane

   (D) nodular hyaline glomerulosclerosis

   (E) plasma cell interstitial nephritis

164. A 73-year-old man is hospitalized with heart failure. The most obvious deleterious effect of a failing heart is the inability to pump enough blood to satisfy the requirements of all the tissues. The patient is given a positive inotropic drug. The greatest benefit derived from administering such a drug to this patient comes from the resulting

   (A) increase in cardiac excitability

   (B) increase in contractile force

   (C) increase in wall thickness

   (D) reduction in heart rate

   (E) reduction in heart size

Questions 165-167: A 56-year-old woman has a paralysis of the right side of her face that produces an expressionless and drooping appearance. She is unable to close her right eye, has difficulty chewing and drinking, perceives sounds as annoying and intense in her right ear, and experiences some pain in her right external auditory meatus. Physical examination reveals loss of blink reflex in the right eye upon stimulation of either cornea and loss of taste from the anterior two-thirds of the tongue on the right side. Lacrimation appears normal in the right eye, the jaw-jerk reflex is normal, and there appears to be no problem with balance.

165. This patient's inability to close her right eye is the result of involvement of

   (A) the buccal branch of the facial nerve

   (B) the buccal branch of the trigeminal nerve

   (C) the levator palpebrae superioris muscle

   (D) the obicularis oculi muscle

   (E) the superior tarsal muscle (of Müller)

166. The difficulty with mastication is the result of paralysis of

   (A) the right buccinator muscle

   (B) the right lateral pterygoid muscle

   (C) the right masseter muscle

   (D) the right zygomaticus major muscle

167. The pain in the external auditory meatus is due to involvement of sensory neurons that have their cell bodies in the

   (A) facial nucleus

   (B) geniculate ganglion

   (C) pterygopalatine ganglion

   (D) spinal nucleus of cranial nerve V

   (E) trigeminal ganglion

168. A 9-year-old girl is brought to the emergency room with the chief complaint of enlarged, painful axillary lymph nodes. The resident physician also notes a small, inflamed, dime-shaped lesion surrounding what appears to be a small scratch on the forearm. The lymph node is aspirated and some pus is sent to the laboratory for examination. A Warthin-Starry silver impregnation stain reveals many highly pleomorphic rod-shaped bacteria. The most likely diagnosis is

   (A) brucellosis
   (B) cat-scratch disease
   (C) plague
   (D) tuberculosis
   (E) yersiniosis

169. Of all cells, those in the central nervous system are most susceptible to ischemia (5 minutes or less), while liver and kidney cells can survive up to 2 h, and epidermal cells will tolerate several hours of hypoxia. In any case, which of the following groups of factors is most important in the cellular pathogenesis of acute ischemia?

   (A) Mitochondrial condensation, glycolysis, sodium cell loss
   (B) Mitochondrial hyperplasia, lysozyme release, membrane injury
   (C) Lipid deposition, reduced protein synthesis, nuclear damage
   (D) Reduced ATP, increased calcium influx, membrane injury
   (E) Ribosome detachment, glycolysis, nuclear damage

170. The Watson-Crick model of DNA structure shows

   (A) a triple-stranded structure
   (B) covalent bonding between bases
   (C) pair bonding between bases A and G
   (D) the DNA strands running in opposite directions
   (E) the phosphate backbone to be on the inside of the DNA helix

171. A 21-year-old female college student went to the college infirmary with the chief complaint of some mild burning during urination and increased frequency of urination. No discharge from the urethra was noted. A "clean-catch" urine specimen was collected and sent to the laboratory for analysis. The urinalysis report returned the next day indicated a "moderate number of leukocytes per high-power field." The bacteriology report read "few *E. coli* (1×10^4 cfu [colony-forming units]/mL)—suspect contamination." This patient most likely has

   (A) anterior urethral syndrome
   (B) gonococcal urethritis
   (C) no disease
   (D) renal disease
   (E) sterile pyuria

172. Most of the energy for muscular contraction is stored in muscle tissue in the form of

   (A) ADP
   (B) ATP
   (C) creatine phosphate
   (D) cyclic AMP
   (E) phosphoenolpyruvate

173. A 69-year-old man is diagnosed with prostatic cancer. On fine needle aspiration of tumor cells it is found to be an adenocarcinoma

(A) in either of the two lateral lobes

(B) in the anterior lobe

(C) in the median lobe

(D) in the posterior lobe

(E) with equal distribution among the lobes

174. Ehlers-Danlos syndrome occurs in several forms. In type IV disease there is a defect in type III collagen synthesis. Which of the following symptoms would be expected in a patient with this disorder?

(A) Hyperextensibility of the integument

(B) Hypermobility of diarthrodial joints

(C) Imperfections in dentin formation

(D) Increased degradation of proteoglycans in articular cartilages

(E) Rupture of the intestinal or aortic walls

175. A 55-year-old man who is being treated for adenocarcinoma of the lung is admitted to a hospital because of a temperature of 38.9°C (102°F), chest pain, and a dry cough. Sputum is collected. Gram stain of the sputum is unremarkable and culture reveals many small gram-negative rods able to grow only on a charcoal yeast extract agar. This organism most likely is

(A) *Chlamydia trachomatis*

(B) *Klebsiella pneumoniae*

(C) *Legionella pneumophila*

(D) *Mycoplasma pneumoniae*

(E) *Staphylococcus aureus*

176. A young man was seen at 2 AM in the emergency room suffering from extreme agitation, a rapid heat rate, and elevated blood pressure. His friends said he had been using cocaine throughout the evening and they were unable to calm him. Which of the following statements about cocaine is true?

(A) It blocks the action of acetylcholine

(B) It is completely hydrolyzed by plasma esterases

(C) It converts cardiac arrhythmias to normal sinus rhythms

(D) It potentiates the effects of injected epinephrine

(E) It produces local vasodilatation

177. A genetic probe for the diagnosis of *Mycoplasma pneumoniae* has recently been discovered. It is rapid (2 h), does not require that DNA be digested to produce single strands, shows little cross-reactivity with other bacteria, and is 100 times more sensitive than other probes. This probe is most likely

(A) a DNA probe that binds to double-stranded DNA

(B) a DNA probe that binds to tRNA

(C) a DNA probe that is specific for the genus *Mycoplasma* and binds to cell-wall constituents

(D) an RNA probe that binds to ribosomal RNA

(E) an RNA probe that binds to mRNA

34

Questions 178-179: At a family picnic in Maine last summer, the following meal was served: baked beans, ham, coleslaw, eclairs, and coffee. Of the 30 people who attended, 4 senior citizens became ill in 3 days; 1 eventually died. Two weeks after attending the picnic, a 24-year-old woman gave birth to a baby who rapidly became ill with meningitis and died in 5 days. Epidemiologic investigation revealed the following percentages of people who consumed the various food items: baked beans 30 percent, ham 80 percent, coleslaw 60 percent, eclairs 100 percent, and coffee 90 percent.

178. All the following statements are true EXCEPT

(A) this is not a case of food poisoning because only 4 people became ill

(B) the death of the baby may be related to the food consumed at the church supper

(C) based on the epidemiologic invest-igation, no one food item can be implicated as the cause of the disease

(D) additional data on the microbiologic analysis of the food are required

(E) additional epidemiologic data should include the percentage of those who ate a particular food item who became ill

179. Microbiologic analysis revealed no growth in the baked beans, ham, or coffee; many gram-positive beta-hemolytic, short, rod-shaped bacteria in the coleslaw; and rare gram-positive cocci in the eclairs. The most likely cause of this outbreak is

(A) *Clostridium botulinum*

(B) *Clostridium perfringens*

(C) *Listeria*

(D) nonmicrobiologic

(E) *Staphylococcus aureus*

---

180. The process of glycolysis includes the following reactions and their concomitant free-energy changes:

Glyceraldehyde 3-phosphate + NAD$^+$ + P$_i$ $\Leftrightarrow$ 1,3-diphosphoglycerate + NADH + H$^+$:
$\Delta G^{\circ\prime}$ = +1.5 kcal/mol

1,3-Diphosphoglycerate + ADP $\Leftrightarrow$ 3-phosphoglycerate + ATP: $\Delta G^{\circ\prime}$ = –4.5 kcal/mol

For the two-step process converting glyceraldehyde 3-phosphate to 3-phosphoglycerate, the overall free-energy change is

(A) $\Delta G^{\circ\prime}$ +6.0 kcal/mol

(B) $\Delta G^{\circ\prime}$ +3.0 kcal/mol

(C) $\Delta G^{\circ\prime}$ –3.0 kcal/mol

(D) $\Delta G^{\circ\prime}$ –4.5 kcal/mol

(E) $\Delta G^{\circ\prime}$ –6.0 kcal/mol

181. Of the following agents, the most effective in treatment of acute migraine headache is

    (A) amitriptyline

    (B) clonidine

    (C) ergotamine tartrate

    (D) methysergide

    (E) propranolol

182. Although malignant melanoma of the skin is not as common as squamous and basal cell carcinoma, it is an exceedingly important and somewhat mysterious tumor owing to its often devastating clinical course and occasionally unpredictable behavior. Which of the following criteria is most significant in predicting the clinical behavior of malignant melanoma on the skin?

    (A) The amount of inflammation

    (B) The degree of pigmentation

    (C) The degree of pleomorphism

    (D) The level of penetration

    (E) The number of halo cells

DIRECTIONS: Each negatively phrased question below contains four or five suggested responses. Select the **one best** response to each question.

183. All the following statements apply to the Pasteur effect EXCEPT that

    (A) adenosine triphosphate (ATP) causes allosteric inhibition of phosphofructo-kinase under aerobic conditions

    (B) 18 times as much glucose per cell is needed to generate the same amount of ATP anaerobically as is generated aerobically

    (C) glycolysis decreases greatly under aerobic conditions

    (D) if oxygen is supplied to an anaerobically grown cell, lactate accumulates

    (E) under aerobic conditions, citrate causes allosteric inhibition

184. Infective endocarditis, unlike rheumatic endocarditis, continues to be a clinical problem even in the antibiotic era, with such new factors as intravenous drug abuse and immunosuppression contributing to its persistence. Acute infective endocarditis differs from subacute endocarditis in all the following respects EXCEPT

    (A) the nature of the causative organisms

    (B) the likelihood of embolization

    (C) the nature of presenting symptoms

    (D) the predisposition to occur on an abnormal cardiac valve

    (E) the likelihood of valve rupture

185. In the fifth week of development, the pancreas develops from a dorsal bud immediately cranial to the hepatic diverticulum and from a ventral bud caudal to, and opposite, the hepatic diverticulum. The dorsal bud contributes to all the following EXCEPT the

(A) accessory pancreatic duct (of Santorini)

(B) body of the pancreas

(C) distal portion of the main pancreatic duct

(D) uncinate process

(E) upper half of the head of the pancreas

186. The chronic drinking of alcohol by a pregnant woman can produce fetal alcohol syndrome. With this syndrome, the newborn may exhibit all the following EXCEPT

(A) decreased chance of normal extra-uterine development

(B) microcephaly

(C) premature birth

(D) reduction in the number of fingers and toes

(E) retardation of intrauterine growth

187. Characteristic histologic features of psoriasis include all the following EXCEPT

(A) acanthosis

(B) epidermal microabscesses containing polymorphonuclear neutrophilic leukocytes

(C) parakeratosis

(D) shortening of the rete ridges and dermal papillae

(E) thin or absent stratum granulosum

188. Binding of a peptide hormone to its receptor may involve all the following EXCEPT

(A) allosteric regulation

(B) coupling with an activity site

(C) formation of covalent linkages

(D) hydrogen bonding

(E) reversible hydrophobic interactions

189. In an unfortunately large number of cases each year, certain children are singled out for abuse by their parents. All the following statements about child abuse are true EXCEPT that

(A) child-abusing parents often were abused by their own parents

(B) fathers tend to abuse their children more often than do mothers

(C) parents are more likely to abuse one "scapegoat" child than all of their children

(D) prematurely born children are more often abused than are normal-term children

(E) younger children are more often abused than are older children

190. A patient is found to have excessive adrenal cortical secretion of aldosterone. Which of the following factors controlling the synthesis and secretion of aldosterone is LEAST important?

(A) Adrenocorticotropic hormone (ACTH)

(B) Angiotensin II

(C) Plasma K^+ concentration

(D) Plasma Na^+ concentration

(E) Renin

191. A health maintenance organization (HMO) is characterized by all the following key principles EXCEPT

   (A) emphasis on efficiency and economy of medical care

   (B) provision of comprehensive medical care to its enrolled members

   (C) provision of financial incentives for the medical staff

   (D) strong emphasis on preventive health services

   (E) voluntary enrollment of members who pay a fixed amount regardless of the services received

192. Catecholamines raise blood glucose levels by all the following means EXCEPT

   (A) inhibition of gluconeogenesis

   (B) inhibition of hepatic glycogen synthetase

   (C) stimulation of cAMP-dependent protein kinase activity

   (D) stimulation of glycogen phosphorylase

193. Because the heart functions at maximal levels of oxygen extraction even at rest, coronary blood flow must increase when myocardial oxygen demand is increased. Coronary blood flow is increased by all the following EXCEPT

   (A) beta-adrenergic blockade

   (B) decrease in arterial P_{O_2}

   (C) decrease in systemic blood pressure

   (D) increase in arterial P_{CO_2}

   (E) vagal stimulation

194. All the following may be required in the construction of a recombinant DNA molecule EXCEPT

   (A) deoxynucleotidyl transferase (terminal transferase)

   (B) DNA ligases

   (C) polynucleotide kinase

   (D) restriction endonucleases

   (E) western blotting

195. You have a patient in your care known to be deficient in glucose-6-phosphate dehydrogenase. This patient may experience hemolysis following treatment with all the following drugs EXCEPT

   (A) dapsone

   (B) chloramphenicol

   (C) primaquine

   (D) pseudoephedrine

   (E) sulfanilamide

196. The CD4+ T cell plays critical roles in all of the following EXCEPT

   (A) activation of macrophages

   (B) induction of B cell function

   (C) induction of natural killer (NK) cell function

   (D) induction of suppressor cell function

   (E) suppression of cytotoxic T cell function

197. Fungi differ from bacteria in all the following ways EXCEPT

   (A) they are susceptible to griseofulvin

   (B) they do not contain peptidoglycan

   (C) they have nuclear membranes

   (D) they lack a plasma membrane

198. The following diagram of the citric acid cycle contains lettered steps where H⁺-e⁻ pairs might be given to the electron-transport chain. Donation of an H⁺-e⁻ pair occurs at all the following steps EXCEPT

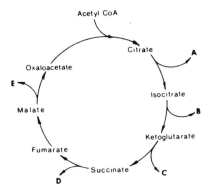

(A) A

(B) B

(C) C

(D) D

(E) E

199. All the following statements correctly describe insulin EXCEPT

(A) it is a small polypeptide composed of two chains connected by disulfide bridges

(B) it is antagonistic to the action of glucagon

(C) it is converted from proinsulin to insulin primarily following secretion from beta cells

(D) it is inactive when in the proinsulin form

(E) it is thought to be an anabolic signal to cells that glucose is abundant

200. All the following statements about the concepts of death held at various stages of child development are true EXCEPT

(A) between birth and 2 years of age, separation is apt to be experienced as synonymous with death

(B) children 3 and 4 years of age consider death to be another form of life

(C) children 5 to 6 years old fantasize that a dead person continues to experience emotion and biological function in the grave

(D) children 7 to 9 years of age realize the inevitability of death, do not feel responsible for the death of others, and yet feel that death can be avoided

(E) at 10 to 12 years of age children continue to feel that death can personally be avoided

201. Testing on patients with so-called "inclusion cell disease" will reveal that all the hydrolytic enzymes are missing from their cells and are found instead in the blood. Inclusion cell disease is a recessive genetic disease where the etiology is related to a cellular sorting malfunction. Each of the following would be expected in inclusion cell disease EXCEPT

(A) absence of a phosphotransferase that adds mannose-6-phosphate groups to lysosomal hydrolases

(B) accumulation of undigested substrates in the cytoplasm

(C) excessive phosphorylation of lysosomal enzymes

(D) mis-sorting of lysosomal enzymes to the secretory pathway

(E) normal expression of the structural genes encoding the hydrolases

202. A 24-year-old woman presents complaining of fever, nausea, headache, chills, and joint pain. You diagnose a viral infection and explain to your patient that her body's immune system will deal with the invading viruses. Each of the following will occur during a viral infection EXCEPT

(A) B cells in the presence of helper T cells and antigen-presenting macrophages differentiate into plasma cells

(B) CD4+ T cells respond to viral antigens and MHC class I molecules to directly attack virus-infected cells

(C) large lymphocytes divide to form plasma cells and memory B cells or cytotoxic T cells and memory T cells

(D) macrophages may phagocytose virus

(E) T- and B-cell areas of the spleen and lymph nodes are involved in the filtration of the blood and lymph, respectively

203. Information that determines the amino acid sequence of proteins is stored within DNA as the so-called "genetic code." All of the following statements regarding the mechanism of information storage are true EXCEPT

(A) a sequence of codons directly corresponds to a sequence of amino acids

(B) codons are the same in all organisms

(C) information is stored as sets of three adjacent bases

(D) more than one codon may exist for a single amino acid

(E) there are 64 different codons that code for amino acids

204. A 17-year-old girl presents with cervical lymphadenopathy, fever, and pharyngitis. Infectious mononucleosis is suspected. Clinically useful tests in this diagnosis include all the following EXCEPT

(A) antibody to Epstein-Barr nuclear antigen (EBNA)

(B) culture

(C) heterophil antibody

(D) IgG antibody to VCA

(E) IgM antibody to viral core antigen (VCA)

205. All the following statements regarding the conversion of ribonucleotides to deoxyribonucleotides are true EXCEPT

(A) the ultimate source of the reducing equivalents is NADPH

(B) the substrates are ribonucleotide diphosphates

(C) the same reductase system is used for all four deoxyribonucleotides

(D) the glycosidic bond between the base and the sugar is cleaved in the process

206. The lung has many metabolic functions in addition to its function in gas exchange. These metabolic functions of the lung include all of the following EXCEPT

(A) activation of angiotensin

(B) inactivation of prostaglandin

(C) synthesis of natriuretic factor

(D) synthesis of prostaglandin

(E) synthesis of surfactant

207. The type of bone shown at low magnification in the photomicrograph below could be found at all the following locations EXCEPT the

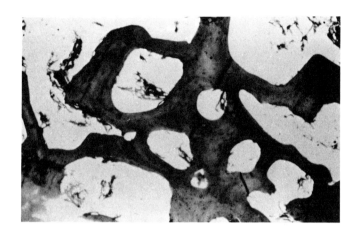

(A) central region (diploe) of flat bones

(B) epiphyses of long bones

(C) medullary cavity of long bones

(D) metaphyses of long bones

(E) outer plate of flat bones

208. Kartagener's syndrome, also known as *immotile cilia syndrome*, may cause all the following EXCEPT

(A) bronchiectasis

(B) immotile stereocilia

(C) infertility in the female

(D) infertility in the male

(E) sinusitis

209. Covalent modification of the enzyme molecule regulates the activity of all the following enzymes EXCEPT

(A) aspartate transcarbamoylase

(B) chymotrypsin

(C) glycogen phosphorylase

(D) glycogen synthetase

(E) pyruvate dehydrogenase

210. A 68-year-old man calls his doctor and complains of blurred vision, nausea, vomiting, and "heart palpitations." He is being medicated with digitalis and his doctor suspects digitalis intoxication. Digitalis intoxication can cause all the following EXCEPT

(A) atrial fibrillation

(B) increased AV conduction

(C) partial atrioventricular block

(D) ventricular extrasystole

(E) ventricular fibrillation

Stop. You have completed this section of the PreTest. Go back over your answers and be sure your answer sheet is carefully marked and that no question has more than one answer. Do not apply any remaining time from this section to another section of this PreTest.

Book B

Time: Three hours

Number of items: 210

Book B

DIRECTIONS: The questions below consist of lettered headings followed by a set of numbered items. For each numbered item select the **one** heading with which it is **most** closely associated. Each lettered heading may be used **once, more than once, or not at all.**

Questions 211-214: Match each numbered structure with its lettered embryonic origin.

    (A) Branchial groove 1

    (B) Branchial arch 1

    (C) Pharyngeal pouch 1

    (D) Branchial groove 2

    (E) Branchial arch 2

    (F) Pharyngeal pouch 2

    (G) Branchial groove 3

    (H) Branchial arch 3

    (I) Pharyngeal pouch 3

    (J) Branchial groove 4

    (K) Branchial arch 4

    (L) Pharyngeal pouch 4

211. Inferior parathyroid gland

212. Stylopharyngeus muscle

213. Stapes

214. Thymus gland

Questions 215-217: For each of the numbered drugs below, select the most suitable lettered description.

    (A) Parenteral penicillin that is resistant to β-lactamase

    (B) Oral penicillin that is resistant to β-lactamase

    (C) Referred to as an extended-spectrum penicillin

    (D) Chemically, the compound is a cephalosporin

    (E) Related to ampicillin but with better oral absorption

    (F) Administered intramuscularly and yields prolonged drug levels

    (G) Cause of a disulfiram-like reaction

    (H) Given parenterally and may cause elevation of serum sodium

    (I) Cause of hypothrombinemia

215. Benzathine penicillin G

216. Methicillin

217. Piperacillin

Questions 218-221: Match each of the numbered descriptions of a hormone below with the appropriate lettered name of that hormone.

    (A) Cortisol

    (B) 1,25-Dihydroxycholecalciferol

    (C) Epinephrine

    (D) Estradiol

    (E) Glucagon

    (F) Insulin

    (G) Luteinizing hormone

    (H) Oxytocin

    (I) Pancreatic polypeptide

    (J) Progesterone

    (K) Prolactin

    (L) Secretin

    (M) Somatostatin

    (N) Thyrotropin-releasing hormone

    (O) Thyroxine

    (P) Vasopressin

218. The D (δ) cells of the pancreas secrete this hormone which functions in the regulation of blood glucose levels

219. A peptide hormone that stimulates milk let-down in the lactating mammary gland

220. Activates phosphorylase in the liver by way of a cyclic nucleotide second messenger mechanism

221. Hormone secreted by the duodenal mucosa, regulates pancreatic exocrine secretions

Questions 222-228: Match each numbered description with the correct lettered cell type.

    (A) Basophil

    (B) Connective tissue mast cell

    (C) Eosinophil

    (D) Fibroblast

    (E) Lymphocyte

    (F) Macrophage

    (G) Mucosal mast cell

    (H) Neutrophil

    (I) Plasma cell

222. Derived from blood or marrow monocytes

223. Recruited through specific basophil and mast cell secretions called *chemoattractant factors*

224. Sole source of blood histamine

225. Microglia and Kupffer cells are examples of this cell type

226. Synthesis of collagen and reticular and elastic fibers as well as the proteoglycans and glycoproteins, which compose the ground substance

227. Presentation of antigen to T and B cells

228. Synthesis of IgE

Questions 229-231: It is customary today to classify antiarrhythmic drugs according to their mechanism of action. This is best defined by intracellular recordings that yield monophasic action potentials. In the accompanying figure, the monophasic action potentials of (A) slow response fiber (SA node) and (B) fast Purkinje fiber are shown. For each numbered description that follows, choose the appropriate lettered drug with which the change in character of the monophasic action potential is likely to be associated.

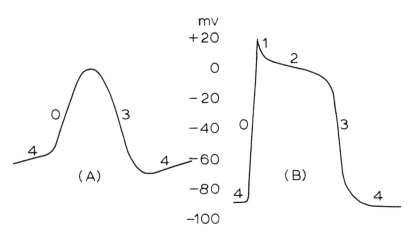

(A) Amiodarone (Cordarone)

(B) Digitalis

(C) Disopyramide (Norpace)

(D) Flecainide (Tambocor)

(E) Mexiletine (Mexitil)

(F) Nifedipine (Procardia)

(G) Propranolol (Inderal)

(H) Verapamil (Calan, Isoptin)

229. Marked phase 0 depression and slow conduction; little effect on repolarization

230. Affects mainly phase 4 of the monophasic action potential of the atrium

231. Minimal phase 0 depression and little slowing of conduction; no effect on or shortens repolarization

---

Questions 232-233: For each numbered disease, choose the lettered sign with which it is most likely to be associated.

(A) Cowdry A intranuclear inclusions

(B) Hepatolenticular degeneration

(C) Lewy bodies

(D) Neurofibrillary tangles

(E) Optic nerve demyelination

(F) Verocay bodies

232. Pick's disease

233. Schwannoma

Questions 234-237: Choose the lettered response that best matches each numbered bacterium.

(A) Secretes erythrogenic toxin that causes the characteristic signs of scarlet fever

(B) Produces toxin that blocks protein synthesis in an infected cell and carries a lytic bacteriophage that produces the genetic information for toxin production

(C) Produces at least one protein toxin consisting of two subunits, A and B, that cause severe spasmodic cough usually in children

(D) Requires cysteine for growth

(E) Secretes exotoxin that has been called "verotoxin" and "Shiga-like toxin"; infection is mediated by specific attachment to mucosal membranes

(F) Possesses *N*-acetylneuraminic acid capsule and adheres to specific tissues by pili found on the bacterial cell surface

(G) Has capsule of polyglutamic acid, which is toxic when injected into rabbits

(H) Synthesizes protein toxin as a result of colonization of vaginal tampons

(I) Causes spontaneous abortion and has tropism for placental tissue due to the presence of erythritol in allantoic and amniotic fluid

(J) Secretes two toxins, A and B, in large bowel during antibiotic therapy

(K) Has 82 polysaccharide capsular types; capsule is antiphagocytic; type 3 capsule (β-D-glucuronic acid polymer) most commonly seen in infected adults

234. *Corynebacterium diphtheriae*

235. *Bordetella pertussis*

236. *Francisella tularensis*

237. *Escherichia coli* 0157/H7

Questions 238-239: Match the following numbered descriptive phrases with the appropriate lettered type of cardiomyopathy.

(A) Constrictive (restrictive) cardiomyopathy

(B) Dilated (congestive) cardiomyopathy

(C) Endomyocardial fibrosis

(D) Hypertrophic cardiomyopathy

(E) Secondary cardiomyopathy

238. Obstruction of left ventricular outflow of blood

239. Usually identified in Southeast Asia and Africa

Questions 240-242: Many families of drugs consist of members that vary only with respect to substituents on a common ring structure. For each numbered type of pharmacologic effect that follows, select the lettered ring structure with which it is most likely to be associated.

240. Bronchodilator

241. Opioid analgesic

242. Antipsychotic

Questions 243-245: Match each numbered vitamin with the appropriate lettered description.

(A) Excess amounts should be avoided when the patient is on levodopa

(B) Overdosage may lead to a psychotic state

(C) Improvement of vision especially in daylight might be attributable to this vitamin

(D) This vitamin is usually not included in the popular "one-a-day" vitamin preparations

(E) Retinoic acid is the natural form

(F) Acute intoxication with this vitamin causes hypertension, nausea and vomiting, and signs of increased CSF pressure

(G) This vitamin has hormonal functions

(H) This fat-soluble vitamin has mainly antioxidant properties

(I) In its water-soluble form, this fat-soluble vitamin is capable of producing kernicterus

243. Pyridoxine

244. Menadione

245. Alpha-tocopherol

Questions 246-248: For each of the following numbered drugs, choose the lettered vitamin, metabolic cofactor, or precursor of which it is a structural analogue.

(A) *p*-Aminobenzoic acid

(B) Calciferol

(C) Folic acid

(D) Nicotinamide

(E) Vitamin K

246. Methotrexate

247. Sulfisoxazole

248. Warfarin

Questions 249-250: Match each numbered clinical situation below with the appropriate lettered risk figure.

(A) 1/10,000

(B) 1/600

(C) 1/100

(D) 1/10

(E) 1

249. The theoretical risk for a 21/21 translocation carrier to have a child with Down syndrome

250. The risk for parents of a trisomy 21 child to have a second offspring with a chromosomal abnormality

Questions 251-252: Match each numbered cytogenetic notation with the appropriate lettered phenotype.

    (A) Down syndrome

    (B) Down syndrome with possible atypical phenotype

    (C) Patau syndrome (the phenotype produced by an extra copy of chromosome 13)

    (D) Translocation carrier with normal phenotype

    (E) Turner syndrome

251. 45,XX,t(21q21q)

252. 46,XX,t(21q21q)

Questions 253-255: For each description below, choose the type of leukemia with which it is most likely to be associated.

    (A) Acute lymphoblastic leukemia

    (B) Acute myeloblastic leukemia (M1)

    (C) Acute promyelocytic leukemia (M3)

    (D) Chronic lymphocytic leukemia

    (E) Hairy cell leukemia

253. Auer rods are frequently present in the leukemic cells

254. It is associated with a short course and diffuse intravascular coagulation

255. It occurs in older adults, produces relatively few symptoms, and is associated with the longest survival

Questions 256-258: For each of the following numbered agents, select the letter of its appropriate site of action in the acetylcholine system shown below.

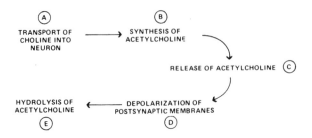

256. Hemicholinium

257. Hexamethonium

258. Muscarine

Questions 259-261: Match the numbered characteristic features with the appropriate lettered acanthotic skin disease.

    (A) Acanthosis nigricans

    (B) Lichen planus

    (C) Pemphigus vulgaris

    (D) Psoriasis

    (E) Verruca vulgaris

259. Elongated, saw-toothed rete ridges and liquefaction degeneration of basal layer

260. Hyperkeratosis, parakeratosis, and vacuolization of the granular layer

261. Papillary, folded hyperkeratosis and melanin pigmentation of basal layer

Questions 262-264: For each of the following numbered patients, select the lettered point on the Frank-Starling curves shown below with which the patient's condition is most likely to be associated. (Point X is the resting state.)

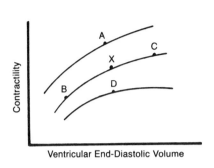

Contractility

Ventricular End-Diastolic Volume

(A) A

(B) B

(C) C

(D) D

262. A 62-year-old man on intravenous saline with dextrose who was mistakenly given an acute volume overload

263. A 26-year-old woman with acute viral pericarditis who developed a pericardial effusion

264. A healthy 15-year-old boy following vigorous exercise

Questions 265-269: Pedigrees I, II, and III in the figure below represent families with retinitis pigmentosa (RP), a genetically heterogeneous eye disease that causes progressive visual impairment. For each numbered situation below, match the appropriate lettered risk.

(A) 50 percent (1/2)

(B) 33 percent (1/3)

(C) 25 percent (1/4)

(D) 11 percent (1/9)

(E) Virtually 0

Pedigree I   **Pedigree II**   **Pedigree III**

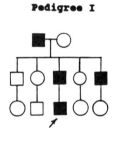
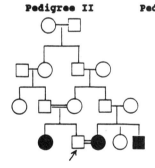
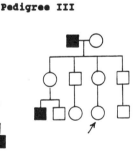

265. For Pedigree I, the risk for proband's son to have RP

266. For Pedigree I, risk for proband's daughter to have RP

267. For Pedigree II, risk for proband's child to have RP

268. For Pedigree II, risk for proband's parents to have another RP child

269. For Pedigree II, risk for proband to have affected child with RP if he had married his wife's unaffected sister

51

Questions 270-272: For each numbered item, choose the lettered growth curve (in an exponentially growing culture) with which it is most likely to be associated. (The arrow in the graph indicates the time at which the drugs were added.)

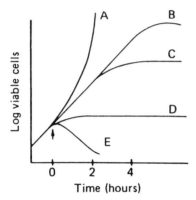

270. Penicillin

271. Sulfonamide

272. Control (without antibiotic)

Questions 273-274: The antigens and antibody in the numbered questions below are associated with hepatitis. For each, choose the lettered description with which it is most likely to be associated.

    (A) Is usually the first viral marker detected in blood after HBV infection

    (B) May be the only detectable serologic marker during the early convalescent phase of an HBV infection ("window phase")

    (C) Appears in the blood soon after infection, rises to very high concentrations, and falls rapidly with the onset of hepatic disease

    (D) Found within the nuclei of infected hepatocytes and not generally in the peripheral circulation

    (E) Closely associated with hepatitis B infectivity and DNA polymerase activity

273. HBcAg

274. Anti-HBc

Questions 275-276: Complement-fixation (CF) testing is an important serologic tool. For each numbered reaction mixture below, select the lettered expected result.

    (A) Complement is bound, red blood cells are lysed

    (B) Complement is bound, red blood cells are not lysed

    (C) Complement is not bound, red blood cells are lysed

    (D) Complement is not bound, red blood cells are not lysed

    (E) Complement is not bound, red blood cells are agglutinated

275. Anti-*Mycoplasma* antibody + complement + hemolysin-sensitized red blood cells (RBC) + anti-RBC antibody

276. Anti-*Mycoplasma* antibody + *Mycoplasma* antigen + complement + hemolysin-sensitized red blood cells

Questions 277-280: For each numbered health care plan described, select the lettered payment plan with which it is most closely associated.

    (A) Blue Cross

    (B) Blue Shield

    (C) Medicaid

    (D) Medicare

    (E) National Health Insurance

277. A compulsory hospital insurance plan whose cost is shared by employees and employers through Social Security payroll taxes

278. A prepayment insurance plan providing limited coverage of hospital costs on a nonprofit basis for individuals or groups

279. A grant-in-aid program for all low-income persons whose costs are shared by the federal and state governments

280. A prepayment supplementary medical insurance program to pay physicians' fees in hospitals, surgery, and emergency care

Questions 281-283: Several major anatomic structures pass through hiatal openings in the diaphragm. Frequently, abdominal contents may herniate into the thorax, either through one of these hiatuses or through developmental defects in the diaphragm. For each description that follows, select the lettered site in the diagram below with which it is most likely associated.

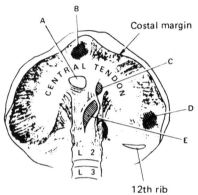

Diaphragm — Inferior Surface

281. The most common site for a diaphragmatic congenital hernia

282. Transmits the right phrenic nerve

283. Transmits a hiatal hernia of the stomach

53

284. A 47-year-old man suffering from mild essential hypertension is being treated with a thiazide diuretic, hydrochlorothiazide. His blood pressure has come down into the normal range and you plan to keep him on the diuretic indefinitely. He should be monitored for which of the following adverse reactions reported with administration of hydrochlorothiazide?

   (A) Hyperchloremia

   (B) Hypermagnesemia

   (C) Hypernatremia

   (D) Hypokalemia

   (E) Hypouricemia

285. The long bacterial structure shown in the electron micrograph below is necessary for

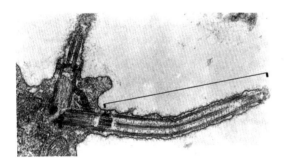

   (A) active transport

   (B) cellular attachment

   (C) cellular rigidity

   (D) conjugation

   (E) motility

286. A 58-year-old hypertensive business executive is being treated with a diuretic. During a regular checkup he is found to be hypokalemic and he has a prolonged PR interval in his electrocardiogram. The PR interval in an electrocardiogram is measured by finding the interval between the

   (A) beginning of the P wave and the beginning of the R wave

   (B) beginning of the P wave and the beginning of the QRS complex

   (C) beginning of the P wave and the end of the QRS complex

   (D) end of the P wave and the beginning of the QRS complex

   (E) end of the P wave and the end of the QRS complex

287. During a physical examination, you discover a young black man has hemoglobin S in his erythrocytes. Which of the following statements regarding hemoglobin S is correct?

   (A) Aggregation of molecules of this form of hemoglobin is favored by high oxygen concentration

   (B) This form of hemoglobin occurs in the erythrocytes of all normal adults

   (C) This form of hemoglobin is characterized by an octapeptide extension at the N-terminus of the beta chain

   (D) A hydrophobic patch on the surface of the molecule causes a tendency to aggregate

   (E) This form of hemoglobin is found most often in South American Indians living at high altitude

288. The piece of tissue shown below (low power view) was probably removed from a

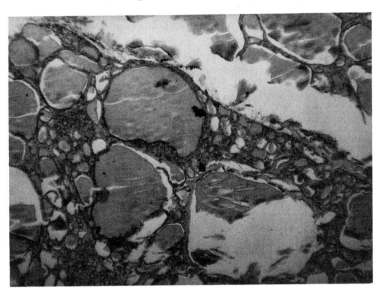

(A) normal thyroid gland

(B) colloid storage goiter

(C) patient with Graves' disease

(D) patient with Riedel's struma

(E) patient with Hashimoto's thyroiditis

---

289. Which of the following conditions will cause an increase in the size of the QRS complex recorded from lead II?

    (A) A shift in the mean electrical axis from +60 degrees to 0 degrees

    (B) Bradycardia

    (C) Second-degree heart block

    (D) Ventricular hypertrophy

    (E) Mitral prolapse

290. The combination of cystic bone lesions, precocious puberty, and patchy skin pigmentations is known as

    (A) Albright's syndrome

    (B) Asherman's syndrome

    (C) Letterer-Siwe disease

    (D) Morquio's disease

    (E) Schaumann's disease

291. The formation of adenosine triphosphate (ATP) is essential for the maintenance of life. In mammalian systems, the number of moles of ATP formed per gram atom of oxygen consumed (the P/O ratio) is 3; in bacteria, however, the P/O ratio may be only 1 or 2. The primary reason for the lower P/O ratio in bacteria is

    (A) absence of nicotinamide adenine dinucleotide (NAD)

    (B) loss of oxidative phosphorylation coupling sites

    (C) less dependence on ATP as an energy source

    (D) absence of a nonphosphorylative bypass reaction

    (E) a less efficient mesosome

292. After receiving incompatible blood, a patient develops a transfusion reaction in the form of back pain, fever, shortness of breath, and hematuria. This type of immunologic reaction is classified as a

(A) complement-mediated cytotoxicity

(B) delayed-type hypersensitivity reaction

(C) systemic anaphylactic reaction

(D) systemic immune complex reaction

(E) T cell-mediated cytotoxicity

293. The terminology, "respiratory control in mitochondria" refers to

(A) the ability of cyanide to inhibit the respiratory chain

(B) the requirement for uncouplers in the synthesis of ATP

(C) the dependence of oxygen consumption upon a process that dissipates the proton-motive force

(D) the independence of calcium transport and the proton-motive force

(E) none of the above

294. A 36-year-old man presented at his physician's office complaining of fever and headache. On examination he had leukopenia and increased liver enzymes and inclusion bodies were seen in his monocytes. History revealed that he was an outdoorsman and remembered removing a tick from his leg. Which of the following diseases should be *excluded* from the differential diagnosis?

(A) Erhlichiosis

(B) Lyme disease

(C) Q fever

(D) Rocky Mountain spotted fever

(E) Tularemia *(Francisella tularensis)*

295. K_m and V_{max} can be determined from the Lineweaver-Burk plot of the Michaelis-Menten equation shown below. When V is the reaction velocity at substrate concentration S, the x-axis experimental data are expressed as

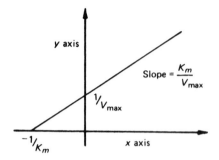

(A) $1/V$

(B) V

(C) $1/S$

(D) S

(E) V/S

296. The pentose phosphate pathway is found in the cytoplasm of all cells. Indicate which of the following statements is true of this pathway

(A) A flavoprotein enzyme catalyzes a key step in the pathway

(B) Carbon dioxide is produced in the nonoxidative portion of the pathway

(C) Glutathione is involved in the interconversion of some of the carbohydrate intermediates in the pathway

(D) Mutational loss of either glucose 6-phosphate dehydrogenase or 6-phosphogluconate dehydrogenase results in an inability to produce pentose phosphates

(E) Mutational loss of glucose 6-phosphate dehydrogenase or 6-phosphogluconate dehydrogenase results in a reduced capacity for the synthesis of NADPH

Questions 297-298: A somewhat obese, 42-year-old mother of three children has experienced several episodes of severe pain in the upper right abdominal quadrant accompanied by pale-colored stools. Though she is not currently experiencing pain, the examiner notes that her skin and sclerae are somewhat yellow. A blood test indicates elevated bilirubin conjugated to glucuronic acid.

297. Bilirubin is produced primarily in which of the following organs?

    (A) Gallbladder

    (B) Kidney

    (C) Pancreas

    (D) Small intestine

    (E) Spleen

298. Elevated bilirubin levels in the blood can result from all the following EXCEPT

    (A) deficiency of an enzyme that makes bilirubin soluble

    (B) hepatocellular damage

    (C) increased destruction of red blood cells

    (D) obstruction of the common bile duct

    (E) obstruction of the cystic duct

---

299. The specimen shown in the photomicrograph below is from a mass removed from the thigh of a 58-year-old man. Using the current nomenclature, this lesion is compatible with

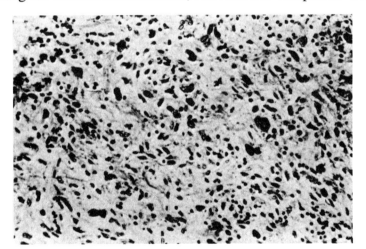

    (A) malignant fibrous histiocytoma

    (B) myositis ossificans

    (C) nodular fasciitis

    (D) osteogenic sarcoma

    (E) rhabdomyosarcoma

300. Which of the following antihypertensive drugs produces most of its effects by blocking α_1-adrenergic receptors in arterioles and venules?

    (A) Clonidine

    (B) Minoxidil

    (C) Phentolamine

    (D) Pindolol

    (E) Prazosin

301. A 6-year-old girl presented to the clinic with scaly patches on the scalp. Primary smears and culture of the skin and hair were negative. A few weeks later, she returned and was found to have inflammatory lesions. The hair fluoresced under Wood's light and primary smears of skin and hair contained septate hyphae. On speaking with the parents, it was discovered that there were several pets in the household. Which of the following is the most likely agent?

    (A) *Epidermophyton floccosum*

    (B) *Microsporum audouini*

    (C) *Microsporum canis*

    (D) *Trichophyton rubrum*

    (E) *Trichophyton tonsurans*

302. A 26-year-old automobile accident victim is on a ventilator adjusted for a tidal volume of 1 L at a frequency of 10/min. If his anatomic dead space is 200 mL and the machine's dead space 50 mL, the alveolar ventilation is

    (A) 10 L/min

    (B) 8.5 L/min

    (C) 7.5 L/min

    (D) 5 L/min

    (E) not determinable from the information given

303. Anterograde amnesia is a specific cognitive deficit in which events that have just occurred are not recalled. It is associated with which of the following disorders?

    (A) Hypochondriasis

    (B) Korsakoff's psychosis

    (C) Manic-depressive psychosis

    (D) Mild retardation

    (E) Sociopathy

304. Two young men became lost while hiking in the high Sierra's. Although water was plentiful, they soon ran out of food. For three days they ate approximately 200 calories per day and then nothing for nearly five days until they finally found their way back to a small village. Initially, their urea excretion was high, but it decreased after a few days of no food. The reason for this is that

    (A) the synthesis of the urea-cycle enzymes decreases

    (B) the use of amino acids for gluconeogenesis decreases

    (C) the use of ketone bodies as fuel decreases

    (D) increased fatty acid synthesis consumes NADPH

    (E) increased protein synthesis uses more amino acids

305. The "inducer" of an inducible enzyme typically interacts with

    (A) the enzyme molecule

    (B) an inhibitor of the enzyme

    (C) a repressor protein

    (D) RNA polymerase

    (E) the DNA coding for the enzyme

306. A 48-year-old man with lymphoma that is being treated with cytotoxic agents has urine, sputum, and blood specimens collected. *Candida albicans* grows from all three. The most effective method to determine whether the *C. albicans* reflects colonization or infection is

- (A) assay of IgA antibody to *C. albicans*
- (B) detection of arabinitol in the blood
- (C) detection of *Candida* protein antigen by latex agglutination
- (D) detection of mannan antigen in the blood
- (E) isolation of *C. albicans* from three body sites

307. In the diagram below of a skeletal muscle myoneural junction, substance 1 is the neurotransmitter substance released by the nerve that stimulates the muscle cell membrane to depolarize. Upon stimulation of the muscle cell membrane, ion 2 enters the muscle cell and ion 3 leaves the cell. Substances 1, 2, and 3 are, respectively,

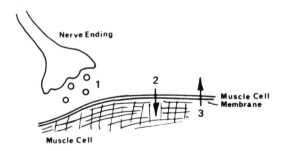

- (A) acetylcholine, chloride, sodium
- (B) acetylcholine, potassium, sodium
- (C) acetylcholine, sodium, potassium
- (D) norepinephrine, sodium, chloride
- (E) norepinephrine, sodium, potassium

308. You see a patient for physical examination. Urinalysis reveals that she is excreting large quantities of methylmalonic acid. This finding could be accounted for by a defect in

- (A) carnitine acyltransferase
- (B) glucose-6-phosphate dehydrogenase
- (C) the absorption of iron
- (D) the absorption of vitamin B_{12}
- (E) vitamin K metabolism

309. An industrial foundry worker who has been chronically exposed to heavy metal vapors has developed a radiographic pattern of pulmonary "honeycombing." Which of the following heavy metals is most likely responsible?

- (A) Arsenic
- (B) Cadmium
- (C) Cobalt
- (D) Lead
- (E) Mercury

310. An AIDS patient with a persistent cough has shown progressive behavioral changes in the past few weeks after eating an undercooked hamburger. A CSF sample is collected and an encapsulated, yeast-like organism is observed. Based only on these observations, what is the most likely organism?

- (A) *Candida*
- (B) *Cryptococcus*
- (C) *Cryptosporidium*
- (D) *Pneumocystis*
- (E) *Toxoplasma*

311. Freeze-fracture is a preparative procedure in which tissue is rapidly frozen and then fractured with a knife. The surfaces or "faces" produced along the fracture planes may be observed under a scanning electron microscope. The face labeled by asterisks in the freeze-fracture preparation shown below may be characterized as

    (A) containing primarily glycoproteins and glycolipids

    (B) facing away from the cytoplasm

    (C) generally possessing a paucity of intramembranous particles

    (D) in direct contact with the cytoplasm

    (E) in direct contact with the external surface

Questions 312-315: A middle-aged man weighing approximately 70 kg and apparently in good health entered a veterans' hospital complaining of severe stomach pains. Following a workup he was diagnosed as suffering from peptic ulcer disease, and cimetidine (Tagamet) 200 mg was administered intravenously. The plasma concentrations of the drug were determined at various times after injection and these values are shown in the figure below.

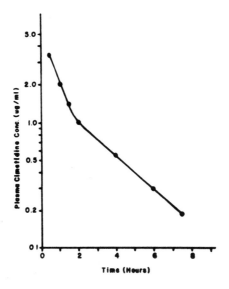

312. The elimination half-life $t_{1/2}$ of cimetidine (Tagamet) in this patient is

(A) 0.4 h

(B) 0.8 h

(C) 1.5 h

(D) 2.3 h

(E) 4.0 h

313. The elimination rate constant (k_e) of cimetidine (Tagamet) in this patient is

(A) $0.1 \ h^{-1}$

(B) $0.2 \ h^{-1}$

(C) $0.3 \ h^{-1}$

(D) $0.4 \ h^{-1}$

(E) $0.5 \ h^{-1}$

314. The apparent volume of distribution of cimetidine (Tagamet) in this patient is

(A) 1.6 L

(B) 39 L

(C) 59 L

(D) 111 L

(E) 200 L

315. The total body clearance of the drug is

(A) 33.5 L/h

(B) 48.0 L/h

(C) 60.0 L/h

(D) 255.0 L/h

(E) 370.5 L/h

316. The diagnosis of fungal infection may be clinical, serologic, microscopic, or cultural. Although the isolation and identification of a fungus from a suspect lesion establishes a precise diagnosis, it is time-consuming. Microscopy is more rapid but generally less sensitive. Visualization of fungi in a clinical specimen is best accomplished by treatment of the specimen with

(A) calcofluor white

(B) hydrochloric acid

(C) para-aminobenzoic acid

(D) potassium hydroxide

(E) silver nitrate

317. Achondroplasia is an autosomal dominant form of skeletal dysplasia that produces dwarfism. Rarely, two affected individuals (heterozygotes) will mate and produce a severely affected homozygote. Representing the abnormal allele frequency by p and the normal allele frequency by q, why does the Hardy-Weinberg law predict that homozygotes will be rare in dominant diseases?

(A) Achondroplasia in homozygotes is prenatally lethal

(B) Affected individuals, represented by the genotype frequency p^2, will be very rare

(C) Affected individuals, represented by the genotype frequency q^2, will be very rare

(D) Assortative mating is rare because achondroplasts do not meet

(E) The genotype frequency $2pq$, which represents affected heterozygotes, will be much larger than p^2, which represents affected homozygotes

318. Two healthy young women with identical tidal volumes and respiratory rates are subjected to spirometry and blood gas measurements. Subject A doubles her tidal volume and decreases her respiratory rate to one-half of baseline. Subject B decreases her tidal volume to one-half of baseline and doubles her respiratory rate. Which of the following statements about the resulting alveolar ventilation in the two women is true?

(A) Alveolar ventilation is unchanged in both subjects

(B) Alveolar ventilation increases in both subjects

(C) Alveolar ventilation decreases in both subjects

(D) Alveolar ventilation increases in subject A and decreases in subject B

(E) Alveolar ventilation decreases in subject A and increases in subject B

319. In the dorsal view of the hand shown below, the lettered areas refer to cutaneous nerve distribution. Trauma to the radial nerve in the arm (brachium) or axilla is most likely to result in loss of sensation over which area?

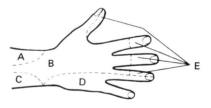

(A) A

(B) B

(C) C

(D) D

(E) E

320. An 82-year-old woman complaining of headaches, visual disturbances, and muscle pain has a biopsy of the temporal artery. The changes revealed by the biopsy specimen are shown in the photomicrograph below. The next course of action is to

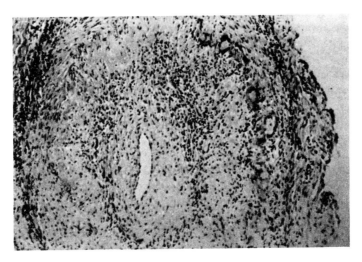

(A) administer anticoagulants

(B) administer corticosteroids

(C) order an erythrocyte sedimentation rate (ESR) test

(D) perform angiography

(E) verify with a repeat biopsy

---

321. According to Freud, the superego may be described as containing the

(A) conscience

(B) pleasure principle

(C) reality principle

(D) primary process thinking

(E) secondary process thinking

322. A patient presents with an adrenal steroid deficiency. Upon appropriate testing, you discover she has a deficiency in steroid 11-hydroxylase. This deficiency, if left untreated, would lead to

(A) adrenal atrophy

(B) goiter

(C) hirsutism

(D) hypernatremia

(E) testicular feminization

323. A 37-year-old woman complains of "stomach cramps," diarrhea, and bloating after she drinks milk or eats more than a relatively small amount of milk products. You determine that she has an intolerance to dietary lactose. Such an intolerance in adults is associated with which one of the following defects?

(A) An inability to hydrolyze lactose in the gut

(B) Lack of the pathway for converting galactose to glucose

(C) An inability to transport galactose across the intestinal lumen

(D) Excessive glucose production when lactose is ingested

(E) Inability to phosphorylate lactose

324. A 60-year-old patient comes in for a complete pulmonary function workup. At the end of a passive expiration, she starts to breathe into a 12-L spirometer containing 10% helium. After several minutes, the helium concentration in the spirometer fell to 8%. This patient's functional residual capacity is approximately

(A) 1 L

(B) 2 L

(C) 3 L

(D) 4 L

(E) 5 L

325. A 7-year-old girl has had repeated infections with *Candida albicans* and respiratory viruses since the time she was 3 months old. As part of the clinical evaluation of her immune status, her responses to routine immunization procedures should be tested. In this evaluation, the use of which of the following vaccines is contraindicated?

(A) BCG

(B) *Bordetella pertussis* vaccine

(C) Diphtheria toxoid

(D) Inactivated polio

(E) Tetanus toxoid

Questions 326-328: While home from college during the December break, a 21-year-old student went to her doctor complaining of malaise, low-grade fever, and a harsh cough, but not of muscle aches and pains. An x-ray revealed a diffuse interstitial pneumonia in the left lobes of the lung. The WBC count was normal. She had been ill for about a week.

326. Based on the information given, the most likely diagnosis is

(A) influenza

(B) legionellosis

(C) *Mycoplasma* pneumonia

(D) pneumococcal pneumonia

(E) staphylococcal pneumonia

327. Based on the information given, which of the following laboratory tests would most rapidly assist you in making the diagnosis?

(A) Cold agglutinins

(B) Complement fixation test

(C) Culture of sputum

(D) Gram stain of sputum

(E) Viral culture

328. The following laboratory data were available within 2 days: cold agglutinins—negative; complement fixation (CF) (*M. pneumoniae*)—1:64; viral culture—pending, but negative to date; bacterial culture of sputum on blood agar and MacConkey's agar—normal oral flora. In order to confirm the diagnosis, which of the following procedures could be ordered to achieve a specific and sensitive diagnosis?

(A) Culture of the sputum on charcoal yeast extract

(B) A repeat cold agglutinin test

(C) A DNA probe to the 16S ribosomal RNA of an organism lacking a cell wall

(D) A repeat CF test in 5 days

(E) Another viral culture in one week

329. An elderly man treated for congestive heart failure for years with digitalis and furosemide dies of pulmonary edema. A postmortem examination of the heart would most likely show

- (A) a dilated, globular heart with thin walls
- (B) aortic and mitral valve stenosis
- (C) right and left ventricular hypertrophy
- (D) right ventricular infarction
- (E) severe left ventricular hypertrophy

330. A 64-year-old cardiac patient is being treated with digitalis in the therapeutic dose range. Her ECG would be likely to show

- (A) elevation of the ST segment
- (B) prolongation of the PR interval
- (C) prolongation of the QT interval
- (D) symmetric peaking of the T wave
- (E) widening of the QRS complex

331. A 24-year-old man runs 10 miles on a moderately warm day. He then drinks 2 L of water to replenish the fluids lost through perspiration during his run. Compared with the situation prior to the run, his

- (A) intracellular fluid will be hypertonic
- (B) extracellular fluid will be hypertonic
- (C) intracellular fluid volume will be greater
- (D) extracellular fluid volume will be greater
- (E) intracellular and extracellular fluid volumes will be unchanged

332. Patient compliance with treatment plans such as the taking of prescribed drugs, special diet, etc., is an important aspect of health care. Interpersonal relationship studies have concluded that the most critical element to assure compliance behavior in a physician-patient relationship is

- (A) allowing the patient to be rewarded in some way for compliance
- (B) congruence of expectations of the physician and the patient
- (C) down-playing of any social class differences between physician and patient
- (D) exchange of accurate information and facts between physician and patient
- (E) similarity of physician's and patient's age

333. A young patient is found comatose in an automobile with the windows up and the engine running. Death ensues within 3 days and an autopsy is performed. Sections through the brain would show

- (A) basilar hemorrhage
- (B) brainstem hemorrhages
- (C) epidural hematoma
- (D) hemorrhages of the lenticular nuclei
- (E) subdural hematoma

334. Which of the following amino acids can give rise to cholesterol, but not to glucose?

- (A) Glycine
- (B) Isoleucine
- (C) Leucine
- (D) Phenylalanine

335. A patient who had extended-wear contact lenses complained to an ophthalmologist about increasing irritation of the eye. The physician sent the patient's contact lens cleaning solution to the laboratory. A wet mount revealed many ameboid organisms. Without further diagnostic or laboratory investigation, the most likely identification of the organism in the lens solution is

    (A) *Acanthamoeba*

    (B) *Hartmannella*

    (C) *Naegleria*

    (D) *Paramecium*

    (E) *Pneumocystis*

336. A 51-year-old man underwent a pulmonary function test. The following data were obtained: fraction of CO_2 in mixed expired gas ($F_E CO_2$) = 3.0%; fraction of CO_2 in alveolar gas ($F_A CO_2$) = 4.5%; tidal volume (V_T) = 450 mL (BTPS); and frequency = 10 breaths per minute. The volume of the physiologic dead space (V_D) is

    (A) 100 mL

    (B) 150 mL

    (C) 225 mL

    (D) 750 mL

    (E) 1500 mL

337. Two subtotal colectomy specimens are sent to the laboratory with both showing a hemorrhagic cobblestone appearance of the mucosa. One, however, shows longitudinal grooving of the surface, which suggests

    (A) Crohn's disease

    (B) ischemic bowel disease

    (C) multiple polyposis syndrome

    (D) ulcerative colitis

338. A deficiency in muscle pyruvate kinase would be expected to result in which of the following symptoms?

    (A) Severe hypoglycemia

    (B) Mental retardation

    (C) Abnormal fatigue upon exercise

    (D) High levels of dihydroxyacetone phosphate and glyceraldehyde 3-phosphate, and low levels of lactate and fructose 1,6-bisphosphate

    (E) Intolerance to fructose

339. A 37-year-old man with pneumonia is being treated with a new penicillin derivative that is partly metabolized and partly excreted unchanged in the urine. Immediately after an initial dose of 100 mg IV, the maximum drug concentration in the plasma was measured at 10 mg/L and the elimination half-life from the patient was estimated to equal 7 h. The plasma clearance of the parent drug appearing in the urine was 8.25 mL/min. What percentage of drug elimination can be attributed to the metabolism of the compound?

    (A) 10

    (B) 25

    (C) 50

    (D) 75

    (E) 90

340. Glucagon secretion results in an elevation of activity of which of the following enzymes?

    (A) Acetyl CoA carboxylase

    (B) Citrate lyase

    (C) Glucose-6-phosphatase

    (D) Glycogen phosphorylase

    (E) Hydroxymethylglutaryl coenzyme A (HMG CoA) reductase

341. The accompanying photomicrograph illustrates which of the following organs?

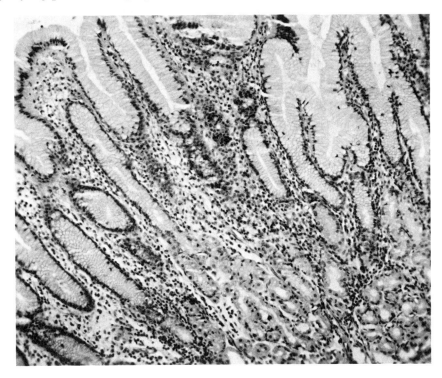

(A) Colon

(B) Esophagus

(C) Fundus

(D) Pylorus

(E) Small intestine

---

342. The sigma factor is an element of RNA polymerase. The role of sigma factor in RNA synthesis is best described by which one of the following statements?

    (A) It separates the strands of double-stranded DNA

    (B) It recognizes specific sites on the DNA template for initiation of RNA synthesis

    (C) It is required for elongation of the RNA molecule

    (D) It interacts with releasing factors to release the completed RNA molecule

    (E) It binds to repressor proteins

343. A young woman succumbed after an 8-month course of severe dyspnea, fatigue, and cyanosis that followed an uneventful delivery of a healthy infant. At necropsy, small atheromas were present in the large and small branches of the pulmonary arteries. Which of the following findings can be predicted in the histologic slides of the lungs?

    (A) Diffuse alveolar hyaline membranes

    (B) Diffuse hemorrhage and infarctions

    (C) Marked medial hypertrophy of pulmonary arterioles

    (D) Multiple pulmonary emboli

    (E) Severe atelectasis and edema

344. A previously healthy 15-year-old boy was admitted to the hospital with severe frontal and bitemporal headache, lethargy, and fever. During the 3 weeks prior to admission, he had been swimming and diving in a freshwater lake. A lumbar puncture was done and examination revealed an elevated white blood cell count, primarily polymorphonuclear leukocytes, and motile amebae. The organism is most likely to be

(A) *Acanthamoeba*

(B) *Dientamoeba fragilis*

(C) *Entamoeba histolytica*

(D) *Entamoeba polecki*

(E) *Naegleria fowleri*

345. A patient hospitalized for fractures of the long bones who develops mental dysfunction, increasing respiratory insufficiency, and renal failure should be suspected of having

(A) aortic valve disease

(B) disseminated intravascular coagulopathy

(C) fat embolism syndrome

(D) myocardial infarction

(E) respiratory distress syndrome

346. An archeologist who recently returned from Saudi Arabia was seen by his family physician for cutaneous lesions on his lips and cheeks. A Giemsa stain of the lesions revealed darkly staining kinetoplasts and light-staining nuclei within macrophages. The most likely cause of these lesions is

(A) *Histoplasma*

(B) *Leishmania*

(C) *Sarcocystis*

(D) *Toxoplasma*

(E) *Trypanosoma*

347. Several drugs are recommended for the lowering of blood cholesterol. Which of the following inhibits the synthesis of cholesterol by blocking 3-hydroxy-3-methylglutaryl-coenzyme A (HMG-CoA) reductase?

(A) Clofibrate

(B) Gemfibrozil

(C) Lovastatin

(D) Nicotinic acid

(E) Probucol

348. Pictured below is an electrocardiogram from a 67-year-old retired school teacher. His cardiac rhythm can be correctly described as

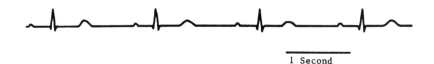

1 Second

(A) atrial flutter

(B) first-degree heart block

(C) second-degree heart block

(D) sinus arrhythmia

(E) tachycardia

349. In the United States, certain enteric protozoan and helminthic infections were previously considered to be exotic illnesses related to foreign travel or to contaminated food or water. However, sexual transmission of these diseases has produced a "hyperendemic" infection rate among male homosexuals. The most common infection seen in this group is

(A) amoebiasis

(B) ascariasis

(C) enterobiasis

(D) giardiasis

(E) trichiuriasis

350. A 32-year-old man breathing at a frequency of 20 breaths per minute has a tidal volume of 400 mL and a dead space volume of 150 mL. His alveolar ventilation is

(A) 250 mL/min

(B) 400 mL/min

(C) 2500 mL/min

(D) 3000 mL/min

(E) 5000 mL/min

351. A patient suffers from severe hypoglycemia that you suspect is due to a genetic defect that has resulted in a deficiency of an enzyme participating in carbohydrate metabolism. A deficiency in which of the following enzymes will result in the most profound hypoglycemia?

(A) Liver amylo-1,6-glucosidase

(B) Liver glucose-6-phosphatase

(C) Liver glycogen phosphorylase

(D) Muscle glucose-6-phosphatase

352. A 25-year-old schoolteacher was well until she attended a church bazaar where she heartily ate barbecued turkey. The following day she developed bloody diarrhea, crampy pain, and tenesmus. A gastroenterologist who did not take a history took a colon biopsy specimen that showed mucosal edema, congestion, and numerous lymphoid cells in the lamina propria. Which of the following differential diagnoses would apply?

(A) Bleeding hemorrhoids vs. Meckel's diverticulitis

(B) Colonic endometriosis vs. amebic dysentery

(C) Early ulcerative colitis vs. *Salmonella* colitis

(D) Staphylococcal gastroenteritis vs. Crohn's disease

(E) Viral gastroenteritis vs. acute diverticulitis

353. You see a 28-year-old woman for a routine Pap smear. While taking a history, you learn that she has an average menstrual cycle of 28 days. However, it has been 35 days since the start of her last menstrual period. A vaginal smear reveals clumps of basophilic cells. You suspect

(A) she will begin menstruating within a few days

(B) she will ovulate within a few days

(C) her serum progesterone levels are very low

(D) there are detectable levels of hCG in her serum and urine

(E) she is undergoing menopause

Questions 354-356: A 66-year-old man, while in the hospital recovering from surgery, is evaluated to determine his acid-base status. The following measurements are taken:

Urine volume = 1.5 L/day

Urinary $[HCO_3^-]$ = 4 meq/L

Urinary titratable
    acids = 10 meq/L

Urinary $[NH_4^+]$ = 20 meq/L

Plasma creatine = 0.8 mg/dL

Urinary creatine = 96 mg/dL

Plasma $[Na^+]$ = 142 meq/L

$[K^+]$ = 3.8 meq/L

Plasma osm = 282 mosm/L

354. What is this patient's daily net acid excretion?

(A) 26 meq/day

(B) 30 meq/day

(C) 34 meq/day

(D) 39 meq/day

(E) 51 meq/day

355. How much new bicarbonate is being formed per day in this patient?

(A) 13 meq/day

(B) 26 meq/day

(C) 34 meq/day

(D) 39 meq/day

(E) 78 meq/day

356. Assuming production of a normal fixed acid load, what do you conclude is this patient's acid-base status?

(A) Cannot be determined from the data

(B) Metabolic acidosis

(C) Normal

(D) Respiratory acidosis

(E) Respiratory alkalosis

357. A middle-aged man has the single presenting symptom of occasional hematuria of very recent onset. The most probable cause is

(A) acute pyelonephritis

(B) mesoblastic nephroma

(C) nephroblastoma

(D) renal cell carcinoma

(E) renal pelvic urothelial tumor

358. Interleukin 1 (IL-1) is a protein with which of the following properties?

(A) It is a lymphocyte-derived product

(B) It may activate B cells

(C) It does not stimulate cytotoxic T cells

(D) One biologically active form is described

(E) Its activity is histocompatibility-restricted

359. A patient presents with sweating and light-headedness after an overnight fast. Blood glucose is low and injections of epinephrine produce a slight increase, but injections of glucagon are not effective in restoring euglycemia. Liver glucose-6-phosphatase activity is normal. The most likely diagnosis is

    (A) liver glycogen phosphorylase deficiency

    (B) muscle glycogen phosphorylase deficiency

    (C) glucagon receptor deficiency

    (D) fructose-1,6-bisphosphatase deficiency

360. A 37-year-old woman is seen upon returning from a 3-week trip to Africa. She complains of having paroxysmal attacks of chills, fever, and sweating. These attacks last a day or two at a time and recur every 36 to 48 h. Examination of a stained blood specimen reveals ring-like and crescent-like forms within red blood cells. The infecting organism most likely is

    (A) *Plasmodium falciparum*

    (B) *Plasmodium vivax*

    (C) *Schistosoma mansoni*

    (D) *Trypanosoma gambiense*

    (E) *Wuchereria bancrofti*

---

361. An excisional biopsy of the nipple area, taken from a 46-year-old woman, is shown below. The patient complained of discharge from the nipple for approximately 4 months. The most likely diagnosis is

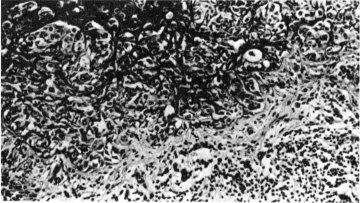

    (A) eczematous inflammation

    (B) epidermoid carcinoma

    (C) fibroadenoma

    (E) Paget's disease of the breast

    (D) mammary fibromatosis

362. A 37-year-old woman presents with a lump in the upper outer quadrant of the left breast, which shows a wide spectrum of benign breast disease on pathologic examination. Which of the following is considered to indicate the greatest risk for subsequent carcinoma of the breast?

(A) Epithelial hyperplasia of the ducts

(B) Florid papillomatosis

(C) Intraductal papillomatosis

(D) Marked apocrine metaplasia

(E) Sclerosing adenosis

363. A 32-year-old woman sees her physician because of "stiffness" and intolerance to cold temperatures in her fingers. Her face has a "masklike" quality. It would be appropriate in the systems review to ask about

(A) family history

(B) headaches and dizziness

(C) sun hypersensitivity

(D) swallowing difficulties

(E) thyroid trouble

364. High serum cholesterol levels are correlated with coronary artery disease. Often dietary reduction of cholesterol intake is sufficient but in some cases, more vigorous treatment with serum cholesterol-lowering drugs is required. Many such drugs designed to lower serum cholesterol are analogues of which of the following compounds?

(A) Acetoacetyl CoA

(B) Acetyl CoA

(C) HMG CoA

(D) Lanosterol

365. A patient presents with adrenoleukodystrophy, a metabolic disorder that results from an absence or deficiency in the peroxisomal enzymes that process long-chain fatty acids. On physical examination of this patient, you would expect to find which of the following signs or symptoms?

(A) Absence of lipid in the brain

(B) Decreased levels of ACTH in the blood

(C) Dementia

(D) Increased production of adrenocortical hormones

(E) Lipid deficiency in the adrenal cortex

366. Acid hydrolysis of a peptide reveals equimolar amounts of lysine, glycine, and alanine. Following trypsin digestion of the peptide, only free glycine and a single dipeptide are observed with paper chromatography. In the following example, the *N*-terminal of the peptide is written to the left of the *C*-terminal end. What is the primary structure of the original peptide?

(A) Ala-Gly-Lys

(B) Ala-Lys-Gly

(C) Gly-Lys-Ala

(D) Gly-Lys-Ala-Lys-Gly-Ala

(E) Lys-Gly-Ala

367. In the developing human embryo, most of the internal organs begin to form in the

(A) first month

(B) second month

(C) third month

(D) fourth month

(E) fifth month

368. The photomicrograph below is of the junction between the

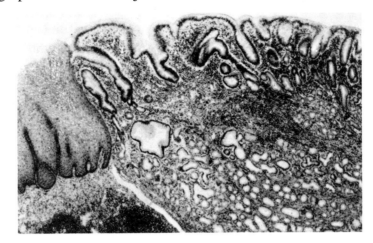

(A) anal canal and rectum

(B) esophagus and stomach

(C) skin of the face and mucous epithelium of the lip

(D) stomach and duodenum

(E) vagina and cervix

---

369. A patient complains of numbness on the left side of the face. On examination, a corneal blink reflex cannot be elicited from the left side, but a normal consensual blink of the left eye is elicited from stimulation of the right cornea. In addition, the patient exhibits deviation of the jaw toward the left during protraction, and loss of pain and temperature senses on the contralateral side of the body and ipsilateral side of the face, as well as coarse intention tremor and a tendency to fall toward the left. There is no contralateral hemiplegia. The most likely location of a brain lesion capable of producing the signs observed in this patient would be the

(A) basal region of the midbrain

(B) lateral region of the mid-pons

(C) medial region of the mid-pons

(D) medial region of the upper medulla

(E) posterolateral region of the caudal medulla

370. You are treating an 82-year-old grandmother and wish to begin her on an angiotensin converting enzyme inhibitor. You should use caution, however, since inhibitors of angiotensin converting enzyme used in elderly persons are apt to cause

(A) marrow depression

(B) drug fever

(C) hepatic injury

(D) renal failure

(E) skin rashes

371. In hereditary fructose intolerance, which of the following enzymes is usually deficient?

(A) Aldolase

(B) Fructose-1,6-bisphosphatase

(C) Hexokinase

(D) Phosphofructokinase

372. A patient complains of muscle weakness of lifetime duration. A muscle biopsy reveals large quantities of fat-filled vacuoles. A homogenate of the biopsy was incubated with a number of uniformly labeled ^{14}C precursors, and labeled CO_2 was collected and counted. It was decided that the patient was deficient in the ability to make carnitine. Which of the following precursors would give labeled CO_2 in assays using this patient's tissue?

(A) Palmitic acid

(B) Palmityl carnitine

(C) Palmityl-CoA

(D) Oleyl-CoA

373. The structure shown below is the parent compound of a group of compounds used primarily as

(A) antihypertensive agents

(B) anti-inflammatory agents

(C) antipsychotic agents

(D) hypnotic agents

(E) tricyclic antidepressants

374. In the case of an inducible gene, the compound called the inducer typically interacts with which of the following?

(A) Adenylate cyclase

(B) A repressor protein

(C) RNA polymerase

(D) The operator region

375. A middle-aged patient is undergoing surgical exploration for a tumor in the pancreatic fundus. No clinical history is given to the pathologist, who notes that the tumor has an "endocrine" appearance in frozen section. An appropriate step in the subsequent evaluation would be to consider

(A) brain scans

(B) celiac angiography

(C) computed tomography

(D) immunofluorescence testing

(E) immunoperoxidase

376. In sickle-cell anemia, red blood cells are distorted in shape and have a tendency to clog capillaries. Patients suffering from sickle cell anemia have an abnormal hemoglobin in their erythrocytes. The abnormality resides in the

(A) addition of heme to hemoglobin

(B) binding of oxygen to heme

(C) incorporation of iron into heme

(D) primary structure of heme

(E) primary structure of one of the globin chains

377. Erik Erikson's stage theory of psychosocial development provides descriptions of personality development that are consistent with a life-span perspective. According to this theory, the stage of young adulthood is best described by a crisis of

(A) autonomy versus dependence

(B) initiative versus guilt

(C) identity versus role confusion

(D) intimacy versus isolation

(E) trust versus mistrust

74

378. Assuming that a variety of microorganisms might cause pneumonia in AIDS patients, which of the following combinations of stains would optimally differentiate the major causes of the pneumonia if a piece of lung tissue was available?

(A) Gram stain and methylene blue stain

(B) Direct fluorescent antibody stain and acid-fast stain

(C) Acid-fast stain and Albert stain

(D) Methylene blue and calcifluor white stain

(E) Lactophenol cotton blue and Giemsa stain

379. A 55-year-old woman is suspected of having a brain tumor because of the onset of seizure activity. Computed tomograms (CT scans) and skull x-rays demonstrate a mass in the right cerebral hemisphere that is markedly calcific. A high index of suspicion should exist for

(A) astrocytoma

(B) brown tumor

(C) cerebral lymphoma

(D) metastatic carcinoma

(E) oligodendroglioma

Questions 380-382: The following measurements are obtained from a 47-year-old female patient during renal function studies.

Plasma		Urine	
Na$^+$	144 mEq/L	flow	2 mL/min
K$^+$	4.1 mEq/L	PAH	750 mL/min
PAH	750 mL/min	creatine	66 mg/dL
creatine	0.8 mg/dL		

380. What is this patient's GFR?

(A) 83 mL/min

(B) 107 mL/min

(C) 165 mL/min

(D) 214 mL/min

(E) Cannot be determined

381. What is the patient's filtration fraction?

(A) 0.18

(B) 0.20

(C) 0.22

(D) 0.24

(E) 0.26

382. Approximately how much glucose is being reabsorbed by this patient's kidneys?

(A) 0 mg/min

(B) 120 mg/min

(C) 165 mg/min

(D) 200 mg/min

(E) 320 mg/min

383. An adult patient with seizure disorder controlled by phenytoin (Dilantin) is noted to have enlarged gingivae. Select the proper course of action below.

(A) No action is necessary because this is a drug effect

(B) No action is necessary because this is a physiologic response

(C) Schilling blood count should be performed

(D) Roentgenograms of the maxillae and mandible should be taken

(E) Biopsy with histologic examination should be performed

Questions 384-385:

384. A 56-year-old woman has a Pa_{CO_2} of 30 mmHg and a plasma bicarbonate concentration of 33 mmol/L. What is her concentration of plasma hydrogen ion (pH)?

(A) 18 nmol/L; pH = 7.75

(B) 28 nmol/L; pH = 7.56

(C) 33 nmol/L; pH = 7.49

(D) 40 nmol/L; pH = 7.40

(E) 48 nmol/L; pH = 7.32

385. The patient from the previous question most likely has which of the following acid-base disorders?

(A) Metabolic acidosis

(B) Metabolic alkalosis

(C) Respiratory acidosis

(D) Respiratory alkalosis

(E) Mixed metabolic alkalosis and respiratory acidosis

386. A 9-year-old girl has been suffering from severe asthmatic attacks. Her pediatrician decides to begin prophylaxic treatment with cromolyn. Cromolyn is believed to exert a beneficial effect because it is

(A) an anticholinergic

(B) a beta$_2$ agonist

(C) a bronchodilator

(D) an H$_1$ receptor blocker

(E) an inhibitor of mediator release

387. The graph below shows the sequential alteration in the type and amount of antibody produced after an immunization. (Inoculation of antigen occurs at two different times, as indicated by the arrows.) Curve A and curve B each represents a distinct type of antibody. The class of immunoglobulin represented by curve B has which of the following characteristics?

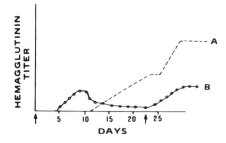

(A) An estimated molecular weight of 150,000

(B) A composition of four peptide chains connected by disulfide links

(C) An appearance in neonates at approximately the third month of life

(D) The human ABO isoagglutinin

(E) A symmetric dipeptide

76

388. The photomicrograph below depicts a biopsy of the uterine cervix that was done following an abnormal Pap smear report. This histologic section shows

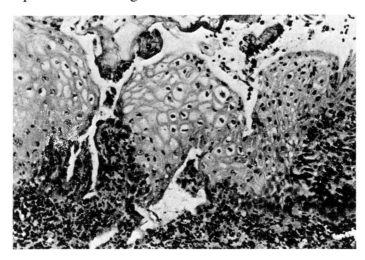

(A) carcinoma in situ

(B) cervical intraepithelial neoplasia

(C) condyloma acuminatum

(D) dysplasia

(E) squamous metaplasia

Questions 389-391: A 53-year-old woman has a paralysis of the right side of her face that produces an expressionless and drooping appearance. She is unable to close her right eye, has difficulty chewing and drinking, perceives sounds as annoying and intense in her right ear, and experiences some pain in her right external auditory meatus. Physical examination reveals loss of blink reflex in the right eye upon stimulation of either cornea and loss of taste from the anterior two-thirds of the tongue on the right side. Lacrimation appears normal in the right eye, the jaw-jerk reflex is normal, and there appears to be no problem with balance.

389. The hyperacusis associated with the right ear results from involvement of

(A) the auditory nerve

(B) the chorda tympani nerve

(C) the stapedius muscle

(D) the tensor tympani muscle

(E) the tympanic nerve of Jacobson

390. The branch of the facial nerve that conveys secretomotor neurons involved in lacrimation is the

(A) chorda tympani

(B) deep petrosal nerve

(C) greater superficial petrosal nerve

(D) lesser superficial petrosal nerve

(E) lacrimal nerve

391. To produce the described signs and symptoms, a lesion involving the facial (CN VII) nerve would be located

(A) in the internal auditory meatus

(B) at the geniculate ganglion

(C) in the facial canal, distal to the geniculate ganglion

(D) at the stylomastoid foramen

(E) within the parenchyma of the parotid gland

392. A 27-year-old man is recovering from a fracture of his right femur suffered in an automobile accident. He has normal pulmonary function: minute volume 4 L; pulmonary blood flow 5 L/min. Suddenly he develops right-sided chest pain and tachypnea. Embolic occlusion of the right pulmonary artery is suspected. The diagnosis would be immediately confirmed by which of the following tracheal gas measurements?

	P_{O_2} (mmHg)	P_{CO_2} (mmHg)
(A)	80	20
(B)	80	60
(C)	100	40
(D)	125	20
(E)	125	60

393. The Arthus reaction is a classic inflammatory response that is best described by which of the following statements?

(A) The Arthus reaction requires a low concentration of antigen and antibody

(B) The Arthus reaction appears later after injection than does passive cutaneous anaphylaxis

(C) The Arthus reaction is mediated by immunoglobulin M

(D) The characteristic Arthus lesion develops slowly

(E) The extent of the Arthus lesion is independent of the quantity of reacting antigen and antibody

DIRECTIONS: Each negatively phrased question below contains four or five suggested responses. Select the **one best** response to each question.

394. The QRS complex of the electrocardiogram represents the depolarization of the ventricles. Which of the following cardiac arrhythmias is LEAST likely to produce a change in the appearance of the QRS complex?

(A) Atrial fibrillation

(B) First-degree heart block

(C) Preventricular contraction

(D) Second-degree heart block

(E) Third-degree heart block

395. Carbon dioxide or bicarbonate is a substrate or product of each of the following enzymes EXCEPT

(A) carboxypeptidase

(B) isocitrate dehydrogenase

(C) phosphoenolpyruvate carboxykinase

(D) pyruvate carboxylase

396. Pulmonary disease sometimes can be localized to a single bronchopulmonary segment, in which case surgical resection may be feasible. Anatomic characteristics of a single bronchopulmonary segment that might assist in its surgical identification include all the following EXCEPT

(A) an apex directed toward the hilum of the lung

(B) a central segmental artery

(C) a central vein

(D) a central tertiary or segmental bronchus

397. By studying erythrocyte cell membranes, much has been learned about membrane function in general. It is known that membrane proteins are able to diffuse laterally within the plane of the lipid bilayer. Each of the following would be expected to limit or reduce that lateral diffusion EXCEPT binding to

(A) actin-based cytoskeleton by proteins such as talin

(B) cell surface antigens by an antibody

(C) fibronectin

(D) laminin

(E) spectrin membrane skeleton elements via ankyrin

398. In humans, ammonia is normally kept below toxic levels by the formation of all the following EXCEPT

(A) carbamoyl phosphate

(B) glutamic acid

(C) glutamine

(D) nitrate

(E) urea

399. A 47-year-old factory worker and smoker of 30 years is being treated for chronic obstructive pulmonary disease (COPD). Caffeine and other methylxanthines are often useful in the treatment of COPD because they have all of the following pharmacologic actions EXCEPT

(A) antagonism of adenosine receptors

(B) constriction of central blood vessels

(C) increased secretion of acid and pepsin by the stomach

(D) relaxation of bronchial smooth muscle

(E) stimulation of cyclic AMP phospho-diesterase

400. The incidence and prevalence of coronary heart disease caused by atherosclerosis have been linked to the type A pattern of coronary-prone behavior. When persons with type-A coronary-prone behavior patterns are subjected to stressful situations, they exhibit increases in all the following physiological responses EXCEPT

(A) cortisol levels

(B) heart rate

(C) plasma norepinephrine levels

(D) occipital alpha wave activity

(E) systolic blood pressure

401. The reduction of oxidized glutathione is central to its role as a sulfhydryl "buffer." All the following statements regarding glutathione are true EXCEPT

(A) it is a tripeptide containing glycine

(B) it contains a glutamic acid residue linked through the gamma carboxyl group

(C) its oxidation by certain antimalarial drugs is normally irreversible

(D) it is pertinent to the manifestation of glucose-6-phosphate dehydrogenase deficiency

(E) its reduced form contains a free sulfhydryl group

402. Which biochemical transformation CANNOT occur in skeletal muscle?

(A) Production of aspartate from oxaloacetate

(B) Production of glucose from glycogen

(C) Production of glycogen from glucose

(D) Utilization of ketone bodies

403. Coenzymes derived from the vitamin shown below are required by enzymes involved in the de novo synthesis of all the following nucleotides EXCEPT

$$\text{H}_2\text{N} \!-\! \text{(pteridine ring, OH)} \!-\! \text{CH}_2 \!-\! \overset{H}{N} \!-\! \text{C}_6\text{H}_4 \!-\! \overset{O}{\text{C}} \!-\! \text{NH} \!-\! \overset{H}{\underset{\text{COOH}}{\text{C}}} \!-\! \text{CH}_2 \!-\! \text{CH}_2 \!-\! \text{COOH}$$

(A) adenosine triphosphate (ATP)

(B) cytidine triphosphate (CTP)

(C) guanosine triphosphate (GTP)

(D) thymidine triphosphate (TTP)

404. All the following statements regarding pyruvate are true EXCEPT

(A) pyruvate kinase catalyzes its reaction with ATP to form phosphoenolpyruvate during hepatic gluconeogenesis

(B) it is not formed by a reversal of the pyruvate dehydrogenase reaction

(C) it is oxidized to acetyl CoA by a multienzyme complex in the mitochondria

(D) it can be made in the cytoplasm from lactate

405. Defects in chemotaxis resulting in increased susceptibility to infection occur in all the following conditions EXCEPT

(A) Chédiak-Higashi syndrome

(B) chronic granulomatous disease of childhood

(C) chronic renal failure of any cause

(D) diabetes mellitus, juvenile type

(E) the neonatal period in full-term infants

406. An adult patient develops crops of bullae and vesicles in the mouth and later on the skin of the trunk. A skin biopsy is inconclusive but shows a suprabasal acantholysis of the overlying epidermis. Direct immunofluorescence of the skin can be used to identify all the following EXCEPT

(A) bullous pemphigoid

(B) dermatitis herpetiformis

(C) discoid lupus erythematosus (DLE)

(D) erythema multiforme

(E) pemphigus vulgaris

407. In addition to their primary structure, proteins are described as having secondary, tertiary, and, sometimes, quaternary structure. All the following are important in the maintenance of the secondary, tertiary, and quaternary structure of proteins EXCEPT

(A) electrostatic interactions

(B) hydrogen bonds

(C) hydrophobic interactions

(D) peptide bonds

(E) van der Waals forces

408. Sickle cell anemia is caused by a point mutation in the hemoglobin gene that results in the substitution of a single amino acid in the mature protein. This mutation could be detected by all the following methods EXCEPT

(A) allele-specific oligonucleotide (ASO) hybridization

(B) DNA sequencing

(C) polymerase chain reaction (PCR) with restriction enzyme digestion

(D) Southern blot analysis

(E) Western blot analysis

409. A 54-year-old man suffering from atrial fibrillation is treated with verapamil. All of the following statements concerning verapamil, a drug that may be used in place of digitalis to treat arrhythmias, are true EXCEPT

(A) it can cause adverse reactions such as constipation and headaches

(B) it has more impact on coronary vascular smooth muscle than on cardiac conduction

(C) it is especially useful in the management of supraventricular arrhythmias

(D) it reduces calcium ion influx through voltage-dependent calcium channels

(E) it undergoes significant first-pass biotransformation following oral administration

410. The elderly use a disproportionate amount of health care resources in the United States. All the following statements concerning health care for the elderly are true EXCEPT that

(A) acute diseases are expected to increase in the elderly over the next 20 years

(B) demands on the health care system are expected to increase for the elderly over the next 20 years

(C) elderly women are heavier users of the health care system than are elderly men

(D) mortality among the elderly has shown a decline over the past 15 years

(E) the sex differential in human longevity is the cumulative result of excessive male mortality throughout the life span

411. All the following statements concerning the U.S. population over the age of 65 years are true EXCEPT

(A) about 5 percent of this population are residents of nursing homes

(B) this population is expected to level off in numbers by the year 2000

(C) this population has been getting larger

(D) the female portion of this population outnumbers the male portion increasingly with advancing age

(E) the number of men per 100 women in the 85-and-older segment is expected to decrease to about 39 by the year 2000

412. Diabetic retinopathy may be characterized by all the following EXCEPT

(A) hemorrhage of retinal vessels

(B) loss of phagocytotic ability of the pigmented epithelium

(C) microaneurysms

(D) retinal ischemia and proliferation of new vessels

(E) thickening of the basal lamina of small retinal vessels

413. All the following statements about citrate are true EXCEPT

(A) it is a positive effector in the synthesis of malonyl CoA

(B) it is a negative effector of phosphofructokinase-1 in liver

(C) it is the source of acetyl CoA for fatty-acid synthesis

(D) it is the source of acetyl CoA in liver mitochondria

414. Primary bile acids are synthesized from cholesterol in the liver. All the following statements about bile acids are correct EXCEPT that they

(A) are absorbed in the intestine and return to the liver via the portal vein

(B) are dehydroxylated by intestinal bacteria

(C) are secreted as conjugated bile salts by the liver

(D) facilitate absorption of fat by emulsifying glycerides

(E) undergo sulfation which promotes their uptake in the intestine

415. Each of the following statements regarding 2,3-diphosphoglycerate (2,3-DPG) is true EXCEPT

(A) it is a product of a side reaction in triglyceride synthesis

(B) it affects the binding of oxygen to hemoglobin

(C) the red blood cell never converts all of its 1,3-DPG into 2,3-DPG

(D) synthesis of 2,3-DPG lowers the ATP yield of glycolysis

416. A known alcoholic is brought to the emergency room following an altercation in a local bar. The intern observes respiratory irregularity, coma, and papilledema. Emergency surgery is planned in order to prevent all the following EXCEPT

(A) brainstem herniation

(B) cerebellar herniation

(C) death of the patient

(D) Duret hemorrhages

(E) ruptured aneurysm

417. A 38-year-old woman, recently divorced, complained of being tired and "run down" for the past few months. On initial screening, her "Monospot" (infectious mononucleosis slide agglutination test) was negative. All the following tests might be helpful in making the diagnosis of chronic infectious mono-nucleosis EXCEPT

(A) EBV-VCA-IgG (EBV viral capsid antigen)

(B) EBV-VCA-IgM

(C) EBV-VCA-IgA

(D) EBNA (EBV nuclear antigen)

(E) EA (early antigen)

418. A healthy, physically fit, 38-year-old male construction worker suffers a cut in an accident at a construction site. His fellow workers estimate that he lost a "quart or two" of blood before a tourniquet could be applied and the bleeding stopped. The hemorrhage this man suffered would be expected to generate all of the following compensatory reactions EXCEPT

(A) arterial vasoconstriction

(B) decreased secretion of catecholamines

(C) decreased venous return

(D) tachycardia

(E) venous vasoconstriction

419. You are treating a 32-year-old male truck driver who has smoked two packs of cigarettes a day for 10 years. He now has cancer that requires chemotherapy. He presents with decreased pulmonary function and his physical examination and chest x-rays suggest a preexisting pulmonary disease. All of the following drugs may be prescribed EXCEPT

    (A) bleomycin (Blenoxane)

    (B) *cis*-diamminedichloroplatinum (Platinol)

    (C) doxorubicin (Adriamycin)

    (D) mithramycin (Mitracin)

    (E) vinblastine (Velban)

420. All the following statements about DNA-primed RNA synthesis are true EXCEPT that

    (A) RNA polymerase catalyzes the formation of phosphodiester bonds only in the presence of DNA

    (B) RNA polymerase requires a primer in the transcription process

    (C) the direction of growth of the RNA chain is from the 5′ to the 3′ end

    (D) only one strand of DNA serves as a template in most circumstances

    (E) the RNA chain synthesized is never circular

Stop. You have completed this section of the PreTest. Go back over your answers and be sure your answer sheet is carefully marked and that no question has more than one answer. Do not apply any remaining time from this section to another section of this PreTest.

Book A

Correct Answers

Number of items: 210

1. **PHYSIOLOGY: ANSWER: G**
2. **PHYSIOLOGY: ANSWER: C**
3. **PHYSIOLOGY: ANSWER: B**

(Berne, 2/e, pp 751-752, 763-764, 769-770, 787-789. West, 12/e, pp 425-426, 433-434, 456, 466-467, 481-482) In the proximal tubule, sodium diffuses passively across the luminal membrane of the epithelial cells and is then actively pumped out of the cell by a Na-K pump located on the basolateral surface. The presence of sodium in the basolateral spaces establishes an osmotic gradient, which causes water to flow out of the lumen. The flow of water down this osmotic gradient maintains the isotonicity of the filtrate and the reabsorbed fluid.

Antidiuretic hormone (ADH) regulates the osmolarity of the extracellular fluid by varying the permeability of the collecting duct to water. The increase in permeability to water caused by ADH allows water to flow out of the collecting duct down an osmotic gradient between the lumen and the medullary interstitium.

The high osmolarity of the medullary interstitium is created by the countercurrent multiplication system of the loop of Henle. The active transport process responsible for the countercurrent multiplier (a carrier that transports one ion of Na^+, one ion of K^+, and two ions of Cl^-) is located on the thick portion of the ascending limb of the loop of Henle.

4. **PHARMACOLOGY: ANSWER: G**

(DiPalma, 3/e, pp 502, 512) Ethinyl estradiol is a synthetic estrogen derivative that is orally effective. It is used in combination with progestins as an oral contraceptive. Ethinyl estradiol is also used alone in various gynecologic disorders such as menopausal symptoms, breast cancer in selected postmenopausal women, and prostatic carcinoma. A major adverse reaction with ethinyl estradiol and other estrogens involves the coagulation reaction. Estrogens increase the synthesis of vitamin K-dependent factors II, VII, IX, and X. The effect on the coagulation scheme can alter the prothrombin time of persons who are using oral anticoagulants (e.g., warfarin). In addition, estrogens can increase the incidence of thromboembolic disorders through their procoagulation effect.

5. **PHARMACOLOGY: ANSWER: E**

(DiPalma, 3/e, pp 502, 512) In the therapy of diabetes mellitus, the effectiveness of insulin to regulate glucose levels in the body can be reduced by simultaneous administration of other drugs. Glucose levels in the body are elevated by the administration of glucocorticoid (e.g., hydrocortisone), dextrothyroxine, epinephrine, thiazide diuretics (e.g., hydrochlorothiazide), and levothyroxine. The drug-induced hyperglycemia counteracts the hypoglycemic action of insulin preparations. In addition, any drug that induces hyperglycemia can also reduce the effectiveness of the oral hypoglycemic agents such as tolbutamide, acetohexamide, and glyburide.

6. **PHARMACOLOGY: ANSWER: L**

(DiPalma, 3/e, pp 502, 512) Spironolactone is classified as a potassium-sparing diuretic. Spironolactone is a competitive inhibitor of aldosterone. It has a mild diuretic effect but is generally used with other diuretics such as thiazides or loop diuretics to prevent the development of hypokalemia. The drug is also used in endocrinology in the diagnosis and treatment of hyperaldosteronism. Another therapeutic use of spironolactone is in the treatment of hirsutism in females, whether it is idiopathic or related to excessive androgen secretion. The drug causes a decrease in the rate of growth and the density of facial hair, possibly through inhibition of excessive androgen production and an effect on the hair follicle.

7. ANATOMY: ANSWER: C
8. ANATOMY: ANSWER: A
9. ANATOMY: ANSWER: H
(Moore, Embryology, 4/e, pp 170-184) See table below.

TABLE OF PRINCIPAL BRANCHIAL DERIVATIVES

	Groove	Arch	Pouch
I	Pinna and external auditory meatus	Mandible, maleus, incus, anterior part of tongue, mm. of mastication, tensors tympani and veli palatini mm., myelohyoid m., ant. belly of digastric m., trigeminal nerve	Auditory tube, middle ear cavity
II		Lesser horns of hyoid, styloid process, stapes, m. of facial expression, stapedius m., stylohyoid m., post. belly of digastric m., facial nerve	Tonsillar fossa
III		Gr. horns of hyoid, post part of tongue, stylopharyngeus m., glossopharyngeal nerve	Vallecular recess. thymus gland, inf. parathyroids
IV		Thyroid cartilage, cricothyroid m., sup. laryngeal nerve	Sup. parathyroids
V			Ultimobranchial bodies (parafollicular cells)
VI		Cricoid and arytenoid cartilages, intrinsic laryngeal mm., inf. laryngeal nerve	Laryngeal ventricle

10. BEHAVIORAL SCIENCE: ANSWER: I

(Brenner, pp 84-100) Reaction formation is the defensive process whereby an unacceptable feeling or impulse is converted into, and consciously experienced as, its opposite. Thus, in reaction formation, "I hate him" becomes "I love him."

11. BEHAVIORAL SCIENCE: ANSWER: H

(Brenner, pp 84-100) Projection is the process through which unacceptable impulses are at once denied and attributed to someone else or to something in the environment. Thus, in projection, "I hate him'' is converted into "he hates me.''

12. BEHAVIORAL SCIENCE: ANSWER: C

(Brenner, pp 84-100) Denial is the defense mechanism by means of which any aspect of reality, e.g., forbidden thoughts, is actively denied. With this defense mechanism, "I hate him" becomes "I don't hate him."

13. PATHOLOGY: ANSWER: A
14. PATHOLOGY: ANSWER: B

(Robbins, 4/e, pp 724-730) Auer rods are often prominent in the hypergranular promyelocytes of acute promyelocytic leukemia since they are formed from the abnormal azurophilic granules. Myeloblasts predominate in acute myeloblastic leukemia (AML) and, therefore, only a few granules, or occasional Auer rods, are present. Acute promyelocytic leukemia is associated with a short course and widespread petechiae and ecchymoses, cutaneous or mucosal, from disseminated intravascular coagulation. The M1 and M3 classes refer to the French-American-British (FAB) classification of AML: M1 is AML in which myeloblasts predominate and M3 is acute promyelocytic leukemia with promyelocytes numerous. Myeloperoxidase is present in both, especially in M3. AML occasionally follows chemotherapy and radiotherapy for Hodgkin's disease.

Chronic lymphocytic leukemia occurs most frequently after the age of 50 (90 percent of cases). It is associated with long survival in many cases and the few symptoms are related to anemia and the absolute lymphocytosis of small mature cells. Splenomegaly may be noted. Some patients are asymptomatic.

Acute lymphoblastic leukemia (ALL) affects children and young adults with marked lymphadenopathy, some splenomegaly, and hepatomegaly. Since chemotherapy at present results in complete remission in 90 percent of children, with more than 50 percent alive 5 years later, it is essential to differentiate ALL from acute myeloblastic leukemia in which prognosis is poor. Cytochemical differentiation includes PAS-positive blasts in most cases of ALL and the presence of terminal deoxynucleotidyl transferase (TdT) in 95 percent of cases of ALL, but in less than 5 percent of acute myeloblastic leukemias.

Hairy cell leukemia and chronic lymphocytic leukemia (CLL) are considered to be chronic lymphoproliferative disorders. CLL is a neoplasm of B cells, like most other lymphoid malignancies. Through molecular analysis, hairy cells are now known to rearrange and express immunoglobulin genes, assigning them also to B-cell lineage.

15. PHARMACOLOGY: ANSWER: C

(DiPalma, 3/e, pp 362, 371. Katzung, 4/e, pp 173-174) It is widely accepted that antiarrhythmic drugs are best classified according to their electrophysiologic attributes. This is best accomplished by relating the effects of the different drugs to their actions on sodium and calcium channels, which are reflected by changes in the monophasic action potential. Disopyramide slows depolarization and repolarization and, like quinidine, delays conduction. Digitalis also affects phase 4 of the action potential, but it also greatly hastens repolarization. Although nifedipine is a calcium channel blocker, it has little effect on the electrophysiology of the heart. Propranolol has actions mainly on slow response fibers and suppresses automaticity.

16. PHARMACOLOGY: ANSWER: A

(DiPalma, 3/e, pp 362, 371. Katzung, 4/e, pp 173-174) Amiodarone blocks sodium channels and markedly prolongs repolarization, particularly in depolarized cells.

17. PHYSIOLOGY: ANSWER: F
18. PHYSIOLOGY: ANSWER: E

A notable exception to the cyclic nucleotide second messenger mechanism is insulin, a polypeptide hormone that acts on a variety of target cells to alter carbohydrate, lipid, and protein metabolism. Although it is clear that insulin exerts its actions by binding to a plasma membrane receptor, the nature of its second messenger is still the subject of intense investigation.

Morphological studies have shown that the cells of the islets are not randomly distributed. The B (β) cells, which secrete *insulin*, are the most abundant and make up the central portion of the islets. These cells are recognized with the electron microscope by the β granules, which are membrane-bound organelles containing several rhomboid insulin crystals. The A (α) cells, which constitute about 20 percent of the cells, secrete *glucagon*. They form a layer around the central core of B (β) cells and are identified ultrastructurally by their α granules, which are membrane-bound organelles with an electron-dense core and peripheral electron-lucent halo. The D (δ) and F cells are also found near the rim of the islets and are in close proximity to both A (α) and B (β) cells. The D (δ) cells secrete *somatostatin* and the F cells secrete pancreatic polypeptide. *Somatostatin* released locally affects the secretion of both the A and B cells, an example of paracrine regulation.

19. GENETICS: ANSWER: B

(Gelehrter, pp 171-189. Thompson, 5/e, pp 201-214) The person in this question has an extra derived chromosome formed by translocation. The extra region includes the 21q21 segment that produces Down syndrome in addition to a duplicated region of chromosome 2 that has been implicated in a different pattern of birth defects.

20. GENETICS: ANSWER: E

(Gelehrter, pp 171-189. Thompson, 5/e, pp 201-214) Deletion of the entire X chromosome or its short arm gives rise to the phenotype of Turner syndrome. Deletion of the X long arm—46,X,i(Xp)—produces a milder phenotype.

21. PATHOLOGY: ANSWER: D

22. PATHOLOGY: ANSWER: A

(Robbins, 4/e, pp 643-648) The cardiomyopathies (CMP) may be classified into primary and secondary forms. The primary forms are mainly idiopathic (unknown cause). The causes of secondary CMP are many and range from alcoholism (probably the most common cause in the United States) to metabolic disorders to toxins and poisons. Whereas there are not many gross organ and microscopic anatomic features of CMP, a few rather characteristic hallmarks are well recognized in separating the types. However, extensive clinical, historical, and laboratory data contribute as much if not more to classification of the type of CMP present than does biopsy or even the postmortem heart examination.

Hypertrophic CMP encompasses those cases in which the major gross abnormality is to be found within the interventricular septum, which is usually thicker than the left ventricle. If there is obstruction of the ventricular outflow tract, there will be moderate hypertrophy in the left ventricles as well, but the septum usually remains thicker, yielding an appearance of asymmetric hypertrophy. This form of CMP occurs in families (rarely sporadically) and is thought to be autosomal dominant. Up to one-third of these patients have been known to die sudden cardiac deaths, often under conditions of physical exertion. Histologically, the myofibers interconnect at angles and are hypertrophied.

Constrictive (restrictive) CMP is associated in the United States with amyloidosis and endocardial fibroelastosis and is so named because the infiltration and deposition of amyloid in the endomyocardium and the layering of collagen and elastin over the endocardium affect the ability of the ventricles to accommodate blood volume during asystole. The heart is more likely to be so involved if the systemic amyloidosis is associated with primary systemic or plasma cell tumors (myeloma). Endocardial fibroelastosis occurs mainly in infants and in the first 1 to 2 years of life and causes a prominent fibroelastic covering to form over the endocardium of the left ventricle. There may be associated aortic coarctation, ventricular septal defects, mitral valve defects, and other abnormalities.

23. HISTOLOGY: ANSWER: C

24. HISTOLOGY: ANSWER: B

25. HISTOLOGY: ANSWER: G

26. HISTOLOGY: ANSWER: H

27. HISTOLOGY: ANSWER: F

28. HISTOLOGY: ANSWER: C

(Junqueira, 7/e, pp 107-120. Roitt, 2/e, pp 19.8-19.12. Ross, 2/e, pp 92-102. Stevens, pp 50-51, 71, 74-77) The cells of connective tissue include resident cells and migrating cells. The developmental source of these cells also varies with some cells derived from undifferentiated mesenchymal cells and others from the marrow. These cell types include fibroblasts, macrophages, mast cells, adipocytes, neutrophils, eosinophils, basophils, plasma cells, and lymphocytes. The fibroblast is the most common connective tissue cell. Fibroblasts are responsible for the synthesis of the fiber (collagen, elastic, and reticular) and ground substance (glycoproteins and proteoglycans) constituents of the connective tissue matrix. Macrophages are phagocytic cells that originate in bone marrow, pass through the bloodstream as monocytes, and ultimately enter tissue. They phagocytose bacteria and viruses, a process initiated by complement and IgG. Macrophages phagocytose antigen and secrete it onto their cell surfaces, where it is presented to other cells, including T and B lymphocytes. In different tissues the phagocytes or macrophages may have different names that reflect an independent discovery by a scientist and therefore an eponym for the cell type (e.g., Kupffer cells in the liver, Langerhans cells in the skin, and Hofbauer cells in the placenta).

Mast cells stain with toluidine blue to form a metachromatic staining product. They are subclassified into two types: (1) the connective tissue mast cell (CTMC), which secretes heparin, is found in the connective tissue and functions independently of T cells; and (2) the mucosal mast cell (MMC), which is located in the mucosa (close to the lumen or surface), secretes chondroitin sulfate rather than heparin, and is dependent on T lymphocytes. Connective tissue mast cells have IgE on their surfaces and secrete histamine, heparin leukotrienes, and chemoattractants called *eosinophil-chemoattractant factors (ECFs)*.

Eosinophils are phagocytes that appear to be specific for antigen-antibody complexes. They also secrete histaminase, which degrades histamine released from basophils and mast cells, and arylsulfatases, which break down leukotrienes. By carrying out these functions, eosinophils modulate the mast cells and basophils that respond during allergic reactions and provide a means of negative feedback regulation of allergic responses. They are attracted to a site of inflammation by ECFs released by mast cells and basophils.

Neutrophils (polymorphonuclear leukocytes) function in the killing of bacteria and aggregate in the area invaded by bacteria. As the cells die, they form one of the major constituents of pus.

29. PHARMACOLOGY: ANSWER: D

(DiPalma, 3/e, pp 414-415, 490-495) Phytonadione, the fat-soluble form of vitamin K, is usually not included in so-called one-a-day vitamin preparations because it is so ubiquitous in the usual diet. Only in liver disease does a deficiency of the vitamin occur.

30. PHARMACOLOGY: ANSWER: G

(DiPalma, 3/e, pp 414-415, 490-495) Calcitriol $(1,25-D_3)$ is the most active form of vitamin D. It is formed by the kidney. When the calcium blood level rises, the kidney produces $24,25-D_3$, a much less active form. Vitamin D can be manufactured in the body by the action of sunlight on the skin. Its main action is to increase calcium absorption in the gut. Thus, vitamin D subserves important hormonal functions in calcium homeostasis.

31. PHARMACOLOGY: ANSWER: L

(DiPalma, 3/e, pp 239, 623-626, 632, 653-656,683-687. Gilman, 8/e, pp 1052, 1055-1056, 1159-1160, 1593-1598) Sulfonamides can cause acute hemolytic anemia. In some patients it may be related to a sensitization phenomenon and in other patients the hemolysis is due to a glucose-6-phosphate dehydrogenase deficiency. Sulfamethoxazole alone or in combination with trimethoprim is used to treat urinary tract infections. The sulfonamide sulfasalazine is employed in the treatment of ulcerative colitis. Dapsone, a drug used in the treatment of leprosy, and primaquine, an antimalarial agent, can produce hemolysis, particularly in patients with a glucose-6-phosphate dehydrogenase deficiency.

32. PHARMACOLOGY: ANSWER: E

(DiPalma, 3/e, pp 239, 623-626, 632, 653-656,683-687. Gilman, 8/e, pp 1052, 1055-1056, 1159-1160, 1593-1598) Ethylene glycol, an industrial solvent and an antifreeze compound, is involved in accidental and intentional poisonings. This compound is initially oxidized by alcohol dehydrogenase and then further biotransformed to oxalic acid and other products. Oxalate crystals are found in various tissues of the body and are excreted by the kidney. Deposition of oxalate crystals in the kidney causes renal toxicity. Ethylene glycol is also a central nervous system depressant. In cases of ethylene glycol poisoning, ethanol is administered to reduce the first step in the biotransformation of ethylene glycol and, thereby, prevent the formation of oxalate and other products.

33. PHARMACOLOGY: ANSWER: G

(DiPalma, 3/e, pp 239, 623-626, 632, 653-656,683-687. Gilman, 8/e, pp 1052, 1055-1056, 1159-1160, 1593-1598) Lead poisoning in children is most often caused by the ingestion of paint chips that contain lead. Older housing units and homes were painted with lead compounds that produced various colors. Chronic lead intoxication causes such symptoms as basophilic stippling, increased δ-aminolevulinic aciduria, tremors, weakness of extensor muscles, constipation, lead line, and colic.

34. PATHOLOGY: ANSWER: C
35. PATHOLOGY: ANSWER: A

(Robbins, 4/e, pp 956, 1397, 1425, 1431, 1445-1446) In classic, or idiopathic, parkinsonism (Parkinson's disease) there is degeneration and loss of pigmented cells in the substantia nigra. With less than 25 percent of nigral cells remaining, there is a deficiency of dopamine in the nigral cells, where dopamine is synthesized, and at the synaptic endings of nigral fibers in the striatum. The severity of the motor syndrome correlates with the degree of dopamine deficiency. Lewy bodies, eosinophilic intracytoplasmic inclusions, are found in remaining neurons of the substantia nigra and are largely localized there in classic Parkinson's disease, unlike diffuse Lewy body disease.

Herpes simplex encephalitis presents a high mortality if not treated with vidarabine (adenine arabinoside, ara-A). It causes necrotizing lesions with Cowdry A intranuclear inclusions in oligodendroglia and occasional neurons. Perivascular mononuclear infiltrates occur and hemorrhagic necrosis is present in the temporal and frontal lobes, especially the orbital gyri.

36. PHARMACOLOGY: ANSWER: E

(DiPalma, 3/e, pp 583-591. Katzung, 4/e, pp 553-558) Amoxicillin is closely related to ampicillin but is better tolerated and has better absorption in the gastrointestinal tract with fewer gastrointestinal side effects. Amoxicillin's spectrum of activity is identical to that of ampicillin, except that ampicillin appears to be more effective against *Shigellosis*.

37. PHARMACOLOGY: ANSWER: H

(DiPalma, 3/e, pp 583-591. Katzung, 4/e, pp 553-558) Carbenicillin is an effective penicillin for the treatment of *Pseudomonas* and *Proteus* infections. It is generally given by the intravenous route but contains 4.7 meq of sodium per gram, which may under chronic use cause an elevation of serum sodium levels. Large doses are often necessary.

38. GENETICS: ANSWER: E
39. GENETICS: ANSWER: B
40. GENETICS: ANSWER: C
41. GENETICS: ANSWER: C
42. GENETICS: ANSWER: D

(Gelehrter, pp 39-44. Thompson, 5/e, pp 72-82) The predominance of affected males with transmission through females makes this pedigree diagnostic of X-linked recessive inheritance. Individual I-1 is an obligate carrier as demonstrated by her affected son and grandson. Individual II-2 cannot transmit an X-linked disorder, although his daughters are obligate carriers. Individual II-3 must be a carrier because of her affected son, which results in a 1 in 4 probability for recurrence of CMT in her offspring. Individual II-5 has a 1 in 2 probability of being a carrier with a 1 in 8 probability for affected offspring. Individual III-4 also has a 1 in 2 probability of being a carrier; her risk for affected offspring is also 1 in 8 despite the consanguineous

marriage. Individual III-8 has a 1 in 4 chance of being a carrier and a 1 in 16 chance of having affected offspring.

43. MICROBIOLOGY: ANSWER: A
44. MICROBIOLOGY: ANSWER: F
45. MICROBIOLOGY: ANSWER: I
(Balows, 5/e, pp 222-258, 277-286, 360-383, 454-456, 471-477. Howard, p 439) Streptococcal infection usually is accompanied by an elevated titer of antibody to some of the enzymes produced by the organism. Among the antigenic substances elaborated by group A β-hemolytic streptococci are erythrogenic toxin, streptodornase (streptococcal DNase), hyaluronidase, and streptolysin O (a hemolysin). Streptolysin S is a nonantigenic hemolysin. Specifically, erythrogenic toxin causes the characteristic rash of scarlet fever.

Many factors play a role in the pathogenesis of *N. meningitidis*. A capsule containing *N*-acetylneuraminic acid is peculiar to *Neisseria* and *E. coli* K1. Fresh isolates carry pili on their surfaces, which function in adhesion. *Neisseria* have a variety of membrane proteins, and their role in pathogenesis can only be speculated upon at this time. The lipopolysaccharide (LPS) of *Neisseria,* more correctly called *lipooligosaccharide* (*LOS*), is the endotoxic component of the cell.

There are no known toxins, hemolysins, or cell wall constituents known to play a role in the pathogenesis of disease by *Brucella*. Rather, the ability of the organisms to survive within the host phagocyte and to inhibit neutrophil degranulation is a major disease-causing factor. In infectious abortion of cattle caused by *Brucella,* the tropism for placenta and the chorion is a function of the presence of erythritol in allantoic and amniotic fluid.

46. PATHOLOGY: ANSWER: D
47. PATHOLOGY: ANSWER: D
(Anderson, 9/e, pp 1757, 1759-1760, 1762-1763, 1772-1774, 1794-1795) Psoriasis occurs in up to 2 percent of the population in Western countries. It is a hereditary skin disease with silver-white, scaling plaques often in sites of repeated trauma. Removal of scale results in minute drops of blood (Auspitz sign). Abnormalities of the nails are found in 25 percent of patients, especially in those with arthritis, and include pitting and yellow-brown "oil spots" under the nail. Histologically, there is marked epidermal thickening and thinning with elongation of rete ridges, parakeratotic hyperkeratosis (nuclei retained in the stratum corneum), and increased mitosis of keratinocytes and other cells above the basal layer. Polymorphonuclear neutrophils (PMNs) form the microabscesses of Munro in the stratum corneum of the epidermis.

48. GENETICS: ANSWER: B
(Gelehrter, chap 8. Thompson, 5/e, chap 9) The incidence of Down syndrome at birth is approximately 1 in 600 liveborn children, with 95 percent trisomy 21.

49. GENETICS: ANSWER: D
(Gelehrter, chap 8. Thompson, 5/e, chap 9) About 4 percent of patients with Down syndrome have translocations that mandate parental karyotyping to determine if one of the parents is a balanced translocation carrier. The remaining 1 percent are mosaics, which means that certain tissues are mixtures of trisomy 21 and normal cells. Translocation carriers have a 5 to 20 percent risk for unbalanced offspring with female carriers in general at higher risk than male carriers.

50. PHARMACOLOGY: ANSWER: C

(DiPalma, 3/e, pp 131-132) Botulinus toxin comes from *Clostridium botulinum*, an organism that causes food poisoning. Botulinus toxin prevents the release of acetylcholine from nerve endings, by mechanisms that are not clear. Death occurs from respiratory failure caused by the inability of diaphragm muscles to contract.

51. PHARMACOLOGY: ANSWER: E

(DiPalma, 3/e, p 68) Isoflurophate is an organophosphate inhibitor of acetylcholinesterase that was developed shortly before World War II. Such inhibitors are widely used as insecticides and, on occasion, in the treatment of glaucoma. Isoflurophate reacts irreversibly with acetylcholinesterase so that the action of acetylcholine cannot be terminated.

52. PHARMACOLOGY: ANSWER: D

(DiPalma, 3/e, p 86) Acetylcholine is synthesized from acetyl-CoA and choline. Choline is taken up into the neurons by an active transport system. Hemicholinium blocks this uptake, depleting cellular choline, so that synthesis of acetylcholine no longer occurs. Tubocurarine is a nondepolarizing agent that binds to the cholinergic receptor at skeletal muscle. It acts as a competitive inhibitor of acetylcholine. Because it is a quaternary ammonium compound, tubocurarine is poorly absorbed from the gastrointestinal tract and is usually given parenterally.

53. MICROBIOLOGY: ANSWER: E
54. MICROBIOLOGY: ANSWER: A

(Jawetz, 19/e, pp 458-462) Advances in the serodiagnosis of viral hepatitis have been dramatic, and the findings of specific viral antigens have led to further elucidation of the course of infections. The "Australian antigen," discovered in 1960, was first renamed hepatitis-associated antigen (HAA) and then, finally, hepatitis B surface antigen (HBsAg). It appears in the blood early after infection, before onset of acute illness, and persists through early convalescence. HBsAg usually disappears within 4 to 6 months after the start of clinical illness except in the case of chronic carriers.

Hepatitis B e antigen (HBeAg) appears during the early acute phase and disappears before HBsAg is gone, although it may persist in the chronic carrier. Persons who are HBeAg-positive have higher titers of HBV and therefore are at a higher risk of transmitting the disease. HBeAg has a high correlation with DNA polymerase activity.

The hepatitis B core antigen (HBcAg) is found within the nuclei of infected hepatocytes and not generally in the peripheral circulation except as an integral component of the Dane particle. The antibody to this antigen, anti-HBc, is present at the beginning of clinical illness. As long as there is ongoing HBV replication, there will be high titers of anti-HBc. During the early convalescent phase of an HBV infection, anti-HBc may be the only detectable serologic marker ("window phase") if HBsAg is negative and anti-HBsAg has not appeared.

55. PHARMACOLOGY: ANSWER: E

(DiPalma, 3/e, pp 96, 121, 224, 255, 267, 301, 309, 539) This compound is cephalexin, an important antimicrobial in the treatment of systemic bacterial infections. The four-membered β-lactam ring found in all the cephalosporin antibiotics is also contained within the structure of the penicillin derivatives.

56. PHARMACOLOGY: ANSWER: A

(DiPalma, 3/e, pp 96, 121, 224, 255, 267, 301, 309, 539) The steroid nucleus, exemplified by prednisone, is shown in Figure **A**. This is the structural basis for androgens, estrogens, progestogens, and anti-inflammatory corticosteroids. Figure **B** is aspirin, a compound that can be categorized as a nonsteroidal anti-inflammatory drug, a nonopioid analgesic, and an antipyretic. Chemically, aspirin belongs to a group of compounds known as salicylates, which includes drugs such as salicylic acid, methylsalicylate, and salsalate.

57. BEHAVIORAL SCIENCE: ANSWER: C

(Goldstein, pp 51-53, 63-64, 102-104, 129-131) The experiment in this question is designed to investigate the possibility that a linear relationship exists between an inhibition constant and the potency of antipsychotic drugs. This is a curve-fitting (as opposed to a difference) hypothesis, and a parametric procedure is justified because the data are continuous, ratio-level variables (linear regression).

58. BEHAVIORAL SCIENCE: ANSWER: B

(Goldstein, pp 51-53, 63-64, 102-104, 129-131) The experiment in this question compares a set of counts in a contingency table with the expected distribution, given the unequal proportions of black and white patients. This is not a two-way analysis of variance, however. The data in the four cells are not means of randomly sampled, ordinal-level, or better variables; they are counts of binary variables (in a cell or not in a cell). The chi-square test is designed for such enumeration data, regardless of the number of categories or the sample size.

59. BIOCHEMISTRY: ANSWER: E

(Stryer, 3/e, p 471) Triacylglycerols are stored when energy and acetyl CoA are available, which fosters the synthesis of fatty acids. These are esterified to glycerol and stored as fat droplets. Glycogen is the storage form of glucose and the other choices are not storage substances.

60. MICROBIOLOGY: ANSWER: C

(Davis, 4/e, pp 705-706) Ornithosis (psittacosis) is caused by *Chlamydia psittaci*. Humans usually contract the disease from infected birds kept as pets or from infected poultry, including poultry in dressing plants. Although ornithosis may be asymptomatic in humans, severe pneumonia can develop. Fortunately, the disease is cured easily with tetracycline.

61. PATHOLOGY: ANSWER: C

(Robbins, 4/e, p 1138) Adenocarcinomas of the vagina and cervix have existed for years, but have increased in young women whose mothers had received diethylstilbestrol (DES) while they had been pregnant. DES was used in the past to terminate an attack of threatened abortion and thereby stabilize the pregnancy. However, a side effect of this therapy proved to be a particular form of adenocarcinoma called clear cell carcinoma. This phenomenon was elucidated by Herpses and Scully in 1970. This unique adenocarcinoma was discovered in daughters between the ages of 15 and 20 of those women who had received DES. The tumor, which carries a poor prognosis, has at least three histologic patterns. One is a tubulopapillary configuration, followed by sheets of clear cells and glands lined by clear cells, and solid areas of relatively undifferentiated cells. Many of the cells have cytoplasm that protrudes into the lumen and produces a "hobnail" (nodular) appearance. Prior to the development of adenocarcinoma, a form of adenosis consisting of glands with clear cytoplasm that resembles that of the endocervix can be seen. This has been termed vaginal adenosis and may be a precursor of clear cell carcinoma. Clinically adenosis of the vagina is manifested by red, moist granules superimposed on the pink-white vaginal mucosa.

62. PHYSIOLOGY: ANSWER: D

(Ganong, 15/e, pp 555-557. Guyton, 8/e, pp 198-199) The carotid sinuses contain baroreceptors that respond to distention by discharging at an increased rate when arterial pressure rises. Impulses from these receptors inhibit the vasomotor center in the brainstem and cause vasodilatation. They also excite the cardioinhibitory center, causing bradycardia. Occlusion of the carotid arteries between the heart and the carotid sinuses would decrease the pressure in the sinus and remove the inhibitory influences on the brainstem, resulting in tachycardia and vasoconstriction with increased arterial pressure.

63. PHYSIOLOGY: ANSWER: B

(Ganong, 14/e, pp 481-485. Guyton, 8/e, pp 106-109) Cardiac output, which is the volume of blood pumped by the heart per minute, is a function of stroke volume (the volume of blood each ventricle ejects per beat) and heart rate (the number of beats per minute). Stroke volume is affected by the load against which blood is pumped (systemic blood pressure). With decreased systemic pressure (caused by a reduction in total peripheral resistance), stroke volume will increase. Contraction of muscle is proportional to muscle length (as described by the Frank-Starling curve), a relationship that holds true even in a denervated heart. Thus increasing end diastolic volume (muscle length) will cause an increase in stroke volume. Sympathetic stimulation increases the strength of contraction at any given length of the cardiac muscle fibers. Increasing the heart rate will usually reduce the end-diastolic volume and thus reduce the stroke volume. However, if the increase in heart rate is caused by an increase in sympathetic stimulation (so that contractility also increases), stroke volume increases.

64. BIOCHEMISTRY: ANSWER: E

(Stryer, 3/e, p 42) Ionizable groups are half ionized at their pK_as. The pH scale is logarithmic with one unit equivalent to a change in ionization of tenfold. The contents of the small intestine is approximately neutral (pH 7). Theophylline, with a pK_a of 8.8, is thus more than 98% nonionized at pH 7.

65. PHARMACOLOGY: ANSWER: C

(DiPalma, 3/e, pp 12-13) Competitive antagonists produce a parallel shift in the dose-response curve of an agonist with no reduction in maximal effect; this is exemplified in the curve shown for norepinephrine plus drug X. Noncompetitive antagonism, as shown with norepinephrine plus drug Y, results in a nonparallel shift in the agonistic dose-response curve and a diminution of the maximum response. Whether X is more potent or more effective than Y as an antagonist cannot be measured by these data.

66. PHARMACOLOGY: ANSWER: B

(DiPalma, 3/e, pp 521-523) Norethindrone is a 19-nortestosterone derivative. This progestational compound possesses a degree of androgenic and anabolic activity. The major use of norethindrone is as an oral contraceptive, alone or in combination with an estrogen. Some adverse reactions attributed to this progestational agent include increased appetite with weight gain, hirsutism, acne, seborrhea, and cholestatic jaundice. The risk of venous thromboembolic disease is associated with estrogenic agents.

67. PHYSIOLOGY: ANSWER: D

(Ganong, 15/e, p 278. Guyton, 8/e, pp 755-756. West, 12/e, pp 744-746) Hormone-sensitive lipase is a cytoplasmic enzyme in adipocytes that catalyzes the complete hydrolysis of triglyceride to fatty acids and glycerol. It is activated by a cyclic AMP-dependent protein kinase that phosphorylates the enzyme, converting it to its active form. Since no accumulation of monoglycerides or diglycerides is detected in adipocytes following the action of hormone-sensitive lipase, it is the initial hydrolysis of triglyceride to fatty

acid and diglyceride that is the rate-limiting step. Hormone-sensitive lipase is sensitive to several hormones *in vitro*, but it appears to be regulated *in vivo* primarily by epinephrine and glucagon, which activate it by increasing cyclic AMP, and insulin, which inhibits it by preventing cyclic AMP-dependent phosphorylation. Cortisol enhances lipolysis indirectly by promoting increased enzyme synthesis.

68. PHYSIOLOGY: ANSWER: B

(Guyton, 8/e, p 413) The T_m is located at the intersection of the extrapolated glucose loss curve and the *x* axis.

69. MICROBIOLOGY: ANSWER: C

(Brown, 5/e, pp 205-212) In the typical life cycle of trematodes, eggs are discharged from the intestinal or genitourinary tract of a definitive host. The eggs hatch in freshwater, releasing the larval miracidia, which enter the snails that serve as intermediate hosts. By metamorphosis, miracidia become rediae, which in turn develop into cercariae. The cercariae are released from the intermediate host and reenter the water. To cause human infection, encysted metacercariae must be ingested; on the other hand, cercariae can penetrate skin. The schizont is an asexual form of malarial protozoa and is not a developmental form of trematodes.

70. PHYSIOLOGY: ANSWER: A

(Berne, 2/e, pp 41-45. Guyton, 8/e, pp 111-117) In order for propagation of an action potential to occur, the depolarization produced by one action potential must depolarize the adjacent patch of excitable membrane to threshold. The amount of charge that must flow to produce this depolarization varies inversely with the membrane capacitance. Thus, as the capacitance increases, the velocity of conduction decreases. The space constant is a measure of how far along the membrane the charge will flow. Velocity of conduction increases as the space constant increases.

71. PHYSIOLOGY: ANSWER: C

(Berne, 2/e, pp 288-296. Guyton, 8/e, p 802) Thermoreceptors in the anterior hypothalamus measure core temperature. Other hypothalamic neurons are involved in the initiation and coordination of heat-conserving and heat-losing mechanisms. Warming the anterior hypothalamic thermoreceptors will lead to activation of heat loss mechanisms including an increase in evaporative heat loss or panting. The other choices all lead to heat conservation.

72. BIOCHEMISTRY: ANSWER: C

(Stryer, 3/e, p 183) The equilibrium constant is independent of the enzyme catalyzing the reaction and is a function only of the concentrations of the products and reactants at equilibrium. There is no reason to expect the other parameters to be comparable.

73. PHARMACOLOGY: ANSWER: A

(DiPalma, 3/e, pp 516-517) Clomiphene (Clomid) is an effective fertility drug that can lead to multiple pregnancies. Clomiphene has been termed an antiestrogen because its stimulant effect on the secretion of pituitary gonadotropins is thought to be the consequence of its blocking the inhibitory effect of estrogens on gonadotropin secretion. Side effects of this drug can include alopecia, breast engorgement, and hot flashes. Oxymetholone is an orally effective anabolic steroid.

74. PATHOLOGY: ANSWER: D

(Robbins, 4/e, pp 133-135) The Barr body represents a sex chromatin clump attached to the nuclear membrane that originates from an entire X chromosome and can easily be seen by using light microscopy to examine scrapings of the epithelium of the inside buccal mucosa. According to the formula $M = n - 1$, the total number of X chromatin masses equals the number of cellular X chromatin masses seen in the nucleus minus 1. Hence, normal males are $0 = 1 - 1$ (no Barr body), and normal females are $1 = 2 - 1$ (one Barr body). In classic Turner's syndrome (XO), the expected buccal smear would be $0 = 1 - 1$ (no Barr bodies seen), as in a normal male. Karyotyping is necessary when the Barr body screening test is ambiguous or inconclusive. In a young woman of short stature and average intelligence who has never menstruated, there is a strong indication that one of the forms of Turner's syndrome exists, and the presence of one Barr body indicates that the patient has XX in some percentage of cells. About 10 percent of all Turner's syndrome patients show a mosaic pattern, with some cells having XO/XX or XO/XXX patterns. In this example, the patient is likely to be XO/XX by the formula $1 = 2 - 1$. In Turner's mosaics, the likelihood of developing a seminoma or gonadoblastoma is higher than expected, and gonadectomy may be indicated.

75. PHARMACOLOGY: ANSWER: B

(Gilman, 8/e, pp 5-7.) As pictured, curve B shows a steady decline in drug level from a maximal initial value and is the pattern found after a bolus intravenous dose. Curve A, reaching and maintaining a constant level, would be obtained by continuous intravenous infusion or by inhalation of air containing a fixed concentration of a drug. Curve C represents plasma levels after an intramuscular injection. Curve D represents the levels obtained from a repository intramuscular injection. Curve E represents levels that occur after a single oral dose. The choice of route for administration of a drug is determined by many factors including serum level desired, sustained level required, reliability of the patient, state of consciousness, the possibility of emesis, and the chemical lability of the drug.

76. ANATOMY: ANSWER: A

(Hollinshead, 4/e, p 659) Visceral afferent pain fibers from the gallbladder travel through the celiac plexus, thence along the greater splanchnic nerves to levels T5-T9 of the spinal cord. Thus, pain originating from the gallbladder will be referred to (appear as if coming from) the dermatomes served by T5-T9, which include a band from the infrascapular region to the epigastrium.

77. MICROBIOLOGY: ANSWER: B

(Howard, pp 273, 279-286) Except during a meningococcal epidemic, *Haemophilus influenzae* is the most common cause of bacterial meningitis in children. The organism is occasionally found to be associated with respiratory tract infections or otitis media.

78. MICROBIOLOGY: ANSWER: A

(Jawetz, 18/e, pp 255-257) Mycoplasmas are extremely small, highly pleomorphic organisms that lack cell walls. They can reproduce on artificial media, forming small colonies with a "fried egg" appearance. They stain poorly with Gram stain but well with Giemsa stain. They are resistant to penicillin but sensitive to tetracycline and sulfonamide.

79. MICROBIOLOGY: ANSWER: A

(Jawetz, 18/e, pp 406-408) Infectious mononucleosis characteristically is accompanied by splenomegaly, the appearance of unique sheep-cell hemagglutinins, an elevated peripheral white blood cell count, and the presence of atypical lymphocytes known as Downey cells. Patients also may develop antibodies to the capsid antigen of Epstein-Barr (EB) virus as measured by immunofluorescent staining of virus-bearing cells.

80. PHYSIOLOGY: ANSWER: A

(Berne, 2/e, pp 613-615. Ganong, 14/e, pp 568-569. Guyton, 8/e, pp 433-437. West, 12/e, pp 540-542) The oxygen-hemoglobin dissociation curve is shifted to the right by increased 2,3-diphosphoglycerate (DPG) levels, increased hydrogen ions, increased CO_2 levels, and a rise in temperature. During exercise, contracting muscle cells release large quantities of carbon dioxide and acids as well as generating heat. All of these effects cause the oxygen-hemoglobin dissociation curve to be shifted to the right and, therefore, the hemoglobin releases oxygen to the muscles. In the lungs, as the partial pressure of CO_2 falls, the pH rises and hemoglobin has an increased affinity for oxygen.

81. PATHOLOGY: ANSWER: C

(Robbins, 4/e, pp 789-791) In diffuse interstitial lung disease (ILD) the early and major event is damage to the alveolar walls. First, an interstitial inflammation affects mainly the septae (interstitial alveolitis) with edema of the alveolar walls and an infiltrate of lymphocytes and monocyte-macrophages. The alveolar lining cells (mostly type I) are injured or become necrotic and are replaced by proliferating type II cells creating a cuboidal epithelial lining; alveolar endothelial cells are also injured, allowing exudation of fluid into the interstitium. If reversal does not occur, the changes become chronic with eventual fibrous scarring of alveolar walls, impaired respiratory function, and pulmonary hypertension.

82. BIOCHEMISTRY: ANSWER: B

(Stryer, 3/e, p 501) Reaction (A) is a simple hydrolysis. Reactions (C) and (D) are freely reversible and unlikely to require the hydrolysis of ATP which is very exothermic. Reaction (B) involves the synthesis of a new chemical bond, which you would expect to require the input of energy.

83. PHARMACOLOGY: ANSWER: D

(DiPalma, 3/e, pp 410-411, 415) The structure shown in the question is hydrochlorothiazide (Esidrix) and is one of several of the thiazide (benzothiadiazide) diuretics. Halogenation of the benzothiadiazine ring at C6 and a free sulfamyl group ($—SO_2NH_2$) at C7 are necessary for maximal diuretic activity in the series of compounds. In contrast to the carbonic anhydrase inhibitors, benzothiadiazides can act independently of acid-base balance. An example of an osmotic diuretic is mannitol, whereas representatives of the loop diuretics are furosemide, ethacrynic acid, and bumetanide. Potassium-sparing diuretics are spironolactone (a steroid), triamterene (a pyrazine derivative), and amiloride (a pyrazinecarbonyl-guanidine).

84. MICROBIOLOGY: ANSWER: A

(Jawetz, 18/e, pp 230-232) Brucellae are small, aerobic, gram-negative coccobacilli. Of the four well-characterized species of *Brucella*, only one—*B. melitensis*—characteristically infects both goats and humans. Brucellosis may be associated with gastrointestinal and neurologic symptoms, lymphadenopathy, splenomegaly, hepatitis, and osteomyelitis.

85. BIOCHEMISTRY: ANSWER: B

(Stryer, 3/e, p 82) Double-stranded DNA displays less optical density than DNA which has been denatured by heat. The term T_m is defined as that temperature at which a given sample of DNA has lost half of its double-helical structure.

86. BIOCHEMISTRY: ANSWER: A

(Stryer, 3/e, p 268) Ascorbic acid (Vitamin C) is derived exclusively from plant sources. Thus, a person eating only meat would quickly become deficient in Vitamin C.

87. MICROBIOLOGY: ANSWER: C

(Jawetz, 18/e, pp 482-485) Parainfluenza viruses are important causes of respiratory diseases in infants and young children. The spectrum of disease caused by these viruses ranges from a mild febrile cold to croup, bronchiolitis, and pneumonia. Parainfluenza viruses contain RNA in a nucleocapsid encased within an envelope derived from the host cell membrane. Infected mammalian cell culture will hemabsorb red blood cells owing to viral hemagglutinin on the surface of the cell.

88. BEHAVIORAL SCIENCE: ANSWER:C

(Conger, 4/e, pp 428-430) College students who do not drink alcohol tend to come from rural, conservative, and deeply religious (usually Protestant) backgrounds. Heavy use of alcohol is found more frequently among social science majors from metropolitan areas, particularly those who are pessimistic about their future. Heavy drinkers are more likely than nondrinkers to use nonalcoholic drugs and have a higher rate of academic failure and drop-out. Their parents are more likely to drink, as well as their friends and best friends.

89. PATHOLOGY: ANSWER: D

(Robbins, 4/e, pp 717-722) Successful treatment of Hodgkin's disease (HD) continues to be the rule, although a few patients have not had the usual dramatic response to therapy. HD is classified as having lymphocyte predominance, lymphocyte depletion, nodular sclerosis, or mixed cellularity according to the histologic appearance. Reed-Sternberg cells, which are the characteristic, large, binucleated cells with prominent nucleoli, are more numerous in the mixed cellularity and nodular sclerosis types and are rare in lymphocyte predominance. Since newer protocols for the treatment of HD even in advanced stages (stages III and IV) are yielding dramatic responses, some workers report their results on overall HD cases without regard to the histologic subtype. The overall prognosis for all HD cases in the United States in 1985 can be said to be relatively very good, and the less common lymphocyte predominance subtype carries an excellent prognosis. The letter "A" attached to the stage number means the patient is asymptomatic and is not anemic, whereas "B" denotes the presence of pruritus, fever, weight loss, and anemia. "Stage I" means that there is lymph node involvement in one region only; "stage II" means that the lymph nodes involved are on the same side of the diaphragm; "stage III" means that both sides of the diaphragm are involved (if the spleen is involved, the letter "s" is affixed); and "stage IV" involves dissemination to extralymphatic tissue, such as marrow, liver, and lung.

90. BEHAVIORAL SCIENCE: ANSWER: D

(Kandel, 3/e, pp 806-810) About 30 to 35 percent of the people who cannot sleep have a relatively simple organic cause for the problem. The two most frequent organic causes are disruptions of normal circadian rhythms and the inevitable consequences of aging. The most common disruptions of normal circadian rhythms are related to travel, "jet lag," and behavioral changes in one's normal daily routine, such as napping, irregular sleep hours and conditions, alteration in meal times, and unusual work schedules. Normal aging is also a major factor as it is more difficult to reset one's biological clock the older one gets. It has been estimated that most people over age 60 sleep only about 5.5 h per day, and since stage 4 NREM sleep also declines with age, the lighter stages of NREM sleep allow the person to awaken more often, sometimes generating the worry that one cannot sleep or that one is not getting enough sleep. Accumulation of hepatic enzymes is most often the result of prolonged use of sleeping pills.

91. PHYSIOLOGY: ANSWER: B

(Berne, 2/e, pp 787-789. Guyton, 8/e, pp 298-302. West, 12/e, pp 464-469) Juxtamedullary nephrons have relatively long loops of Henle that extend into the hypertonic medullary pyramids and constitute about 15

percent of the total nephrons in the human. Glomerular filtrate enters the loop of Henle isotonic to plasma. Because of the active extrusion of sodium, the fluid leaves the loop hypotonic to plasma. It then equilibrates with the cortical interstitial fluid and enters the collecting duct isotonic to plasma. Equilibration continues as the fluid passes through the hypertonic medullary countercurrent mechanism. The resulting urine is hypertonic to plasma.

92. PHARMACOLOGY: ANSWER: C

(DiPalma, 3/e, p 356) Quinidine is often given in conjunction with digitalis. It has been found by pharmacokinetic studies that this combination results in quinidine's replacing digitalis in tissue binding sites (mainly muscle), thus raising the blood level of digitalis and decreasing its volume of distribution. A mechanism by which quinidine interferes with the renal excretion of digitalis has also been proposed.

93. PHYSIOLOGY: ANSWER: A

(Ganong, 15/e, p 63. Guyton, 8/e, pp 74-78) The single muscle cell generates only a single, sudden contraction or twitch that is "all or none" for that cell. During summation, individual muscle twitches are added together to make strong muscle movements. Indeed, the tension developed during summation is much greater than during the single muscle twitch. When a muscle is stimulated at progressively greater frequencies, activation of the contractile mechanism occurs repeatedly before any relaxation has occurred and the successive contractions fuse into one continuous contraction. Such a response is called tetanus. Complete tetanus means that there is no relaxation between stimuli; during incomplete tetanus there are periods of incomplete relaxation between the summated stimuli. During complete tetanus, the tension developed is about four times that developed by the individual twitch contractions.

94. PATHOLOGY: ANSWER: E

(Anderson, 9/e, pp 2187-2188) Subdural hematomas are most commonly located over the cerebral hemisphere convexities. The traditional explanation for the formation of subdural hematomas has been tearing of the bridging veins that pass from the cortical surface to the superior sagittal sinus. Blood may also leak from lacerated cortical vessels or arachnoidal vessels ruptured by a meningeal tear.

95. MICROBIOLOGY: ANSWER: C

(Jawetz, 18/e, pp 25-27) Dipicolinic acid, formed in the synthesis of diaminopimelate (DAP), is a prominent component of bacterial spores but is not found in vegetative cells or in eukaryotic appendages. The calcium salt of dipicolinic acid apparently plays an important role in stabilizing spore proteins, but its mechanism of action is unknown. Dipicolinic acid synthetase is an enzyme unique to bacterial spores.

96. BIOCHEMISTRY: ANSWER: C

(Stryer, 3/e, p 513) Phenylketonuria results from a lack of phenylalanine hydroxylase, which prevents the conversion of phenylalanine to tyrosine, the product of normal degradation. Phenylalanine is transaminated to phenylpyruvate, which accumulates and interferes with brain development by an as yet unknown process.

97. PHARMACOLOGY: ANSWER: E

(DiPalma, 3/e, p 516) Tamoxifen is an estrogen antagonist used in the treatment of certain breast cancers. Postmenopausal women with metastases to soft tissue and whose tumors contain estrogen receptors are most likely to respond to this agent. Little or no benefit is derived from tamoxifen if the tumor does not have estrogen receptors.

98. PATHOLOGY: ANSWER: A

(Robbins, 4/e, pp 1033-1041) While many varieties of glomerulonephritis can produce the nephrotic syndrome, a few disorders will virtually always produce it. Included in the latter group are focal (segmental) glomerulosclerosis, membranous glomerulonephritis (GN), lipoid nephrosis, membranoproliferative glomerulonephritis, systemic diseases (such as amyloidosis and systemic lupus erythematosus), some tumors, hepatitis B, syphilis, drugs such as penicillamine, and certain allergies. Light microscopy shows very little change in glomeruli in lipoid nephrosis, and a diffuse absence of glomerular epithelial foot processes is noted with electron microscopy. Membranoproliferative GN is characterized by an increase in mesangial cellularity accompanied by splitting of the glomerular basement membranes ("double contour"). Membranous GN shows electron-dense deposits of immunoglobulin in the subepithelial portion of the basement membrane. The nephrotic syndrome includes massive albuminuria with significant loss of protein (more than 3 to 5 g of protein) in 24 h, consequent reduced plasma albumin (less than 3 g/dl), hyperlipidemia, and anasarca (generalized edema).

99. MICROBIOLOGY: ANSWER: A

(Jawetz, 18/e, pp 307-309) Patients with disseminated candidiasis must be treated with amphotericin B or other antifungal drugs. Nystatin, the treatment for candidiasis of the mouth (thrush), does not concentrate in tissues and thus is ineffective in treating disseminated candidiasis.

100. MICROBIOLOGY: ANSWER: A

(Lannette, 4/e, pp 461-471) The patient presented with typical symptoms of actinomycosis. *Actinomyces israelii* is normal flora in the mouth. However, it causes a chronic draining infection, often around the maxilla or the mandible, with osteomyelitic changes. Treatment is high-dose penicillin for 4 to 6 weeks. The diagnosis of actinomycosis is often complicated by the failure of *A. israelii* to grow from the clinical specimen. It is an obligate anaerobe. Fluorescent antibody (FA) reagents are available for direct staining of *A. israelii*. A rapid diagnosis can be made from the pus. FA conjugates are also available for *A. viscosus* and *A. odontolyticus*, anaerobic actinomycetes that are rarely involved in actinomycotic abscesses.

101. PHYSIOLOGY: ANSWER: A

(Ganong, 15/e, pp 548-549. Guyton, 8/e, pp 164-167) Normal right atrial pressure is approximately 5 mmHg. Because of hydrostatic pressure, in the standing position venous pressure in the legs is greater than central venous pressure, whereas pressure in the head and neck is less. In the neck, normal veins collapse when a person is in an upright position; intracranially, the dural sinuses, having more rigid walls, are unable to collapse and sinus pressure may drop below 0 mmHg. During an intracranial operation upon a patient in a sitting position, the resultant negative pressure presents a danger that atmospheric air may be drawn into an exposed sinus, causing an air embolus.

102. BIOCHEMISTRY: ANSWER: C

(Stryer, 3/e, p 353) Actively contracting muscle uses too much glucose to accomodate all of the resulting pyruvate within its own metabolism. Excess pyruvate is reduced to lactate which then enters the blood. Upon reaching the liver, this lactate is oxidized to pyruvate which is used for gluconeogenesis. This whole process is known as the Cori cycle.

103. PATHOLOGY: ANSWER: C

(Robbins, 4/e, pp 1031-1032) In Goodpasture's syndrome, circulating antibodies reactive with the glomerular basement membrane will bind in a linear pattern along the entire length of the glomerular basement membrane, which is their specific antigen. IgG is deposited in the basement membrane, along with complement. There are focal interruptions of the glomerular basement membrane as well, along with deposits of fibrin, as seen with electron microscopy.

104. BIOCHEMISTRY: ANSWER: D

(Stryer, 3/e, p 582) Homocysteine and tetrahydrofolate have no free methyl group. Methionine has no tendency to donate its methyl group until it is activated by ATP to form *S*-adenosylmethionine, the methyl donor in a large number of reactions. N^5-methyltetrahydrofolate donates its methyl group only to homocysteine, forming methionine.

105. HISTOLOGY: ANSWER: C

(Junqueira, 7/e, pp 206-207. Ross, 2/e, p 205) The histologic sample must be skeletal or cardiac muscle because of the presence of cross-striations. The presence of peripherally placed nuclei eliminates cardiac muscle as a possible tissue. Skeletal muscle may be subclassified into three muscle fiber types. Red muscle fibers have a high content of cytochrome and myoglobin (an oxygen storage pigment, analogous to hemoglobin found in red blood cells). Red muscle contains many mitochondria beneath the myofiber cell membrane that function in the high metabolism of these cells. Mitochondria are also found in a longitudinal array surrounding the myofibrils. The presence of numerous mitochondria provides a strong staining reaction with the use of cytochemical stains such as that for succinic dehydrogenase. Physiologically, red fibers are capable of continuous contraction but are incapable of rapid contraction. White muscle fibers would stain very lightly for succinic dehydrogenase and there would be few mitochondria visible at the ultrastructural level. These would be primarily associated with the triads as occurs in all forms of skeletal muscle. White fibers are capable of rapid contraction but are unable to sustain continuous heavy work. They are larger than red fibers and have more prominent innervation and synaptic vesicles. Human skeletal muscle fibers are composed of red, white, and intermediate type fibers. The intermediate fibers possess characteristics of both red and white fibers including a size and innervation pattern intermediate between red and white muscle fibers.

106. PATHOLOGY: ANSWER: D

(Sabiston, pp 586-588.) This histology is classic for Paget's disease. The nipple may be scaly or moist and oozing, and an underlying carcinoma will virtually always be found.

107. ANATOMY: ANSWER: A

(Hollinshead, 4/e, pp 876-878) A mnemonic device for remembering the order in which the soft tissues overlie the cranium is SCALP: Skin, Connective tissue, Aponeurosis, Loose connective tissue, and Periosteum. The scalp proper is composed of the outer three layers, of which the connective tissue contains one of the richest cutaneous blood supplies of the body. The occipitofrontal muscle complex inserts into the epicranial aponeurosis, which forms the intermediate tendon of this digastric muscle. This structure, along with the underlying layer of loose connective tissue, accounts for the high degree of mobility of the scalp over the pericranium. If the aponeurosis is lacerated transversely, traction from the muscle bellies will cause considerable gaping of the wound. Secondary to trauma or infection, blood or pus may accumulate subjacent to the epicranial aponeurosis.

108. PHYSIOLOGY: ANSWER: C

(Ganong, 15/e, pp 616-618. Guyton, 8/e, pp 436-437) The hemoglobin-oxygen dissociation curve is sigmoidal. Normally, P_{O_2} and hemoglobin saturation lie at the top of the curve so that hemoglobin saturation varies over a narrow range. Under physiologic conditions, the P_{O_2} of arterial blood is 100 mmHg and hemoglobin is approximately 98 percent saturated with oxygen. Some venous blood bypasses the lungs and prevents saturation from reaching 100 percent. Blood entering the right ventricle represents venous blood where P_{O_2} is at its lowest (40 mmHg) and hemoglobin saturation reaches its minimum of 75 percent.

109. MICROBIOLOGY: ANSWER: E

(Jawetz, 18/e, pp 500-505) The definitive diagnosis of rabies in humans is based on the finding of Negri bodies, which are cytoplasmic inclusions in the nerve cells of the spinal cord and brain, especially in the hippocampus. Negri bodies are eosinophilic and generally spherical in shape; several may appear in a given cell. Negri bodies, although pathognomonic for rabies, are not found in all cases of the disease.

110. BIOCHEMISTRY: ANSWER: B

(Stryer, 3/e, p 874) Reverse transcriptase is the enzyme used by RNA-containing viruses, such as HIV, to make a DNA copy of its genome that can integrate with the host's DNA. It is named for the flow of genetic information from the level of RNA to that of DNA, rather than the other way around.

111. PHARMACOLOGY: ANSWER: E

(DiPalma, 3/e, pp 618-619) Tetracyclines, as chelating agents, have a high affinity for the divalent cations of calcium and magnesium salts, for iron-containing preparations, for dairy products, and for aluminum hydroxide gels. The chelated complex is insoluble and not absorbed through the mucosa of the gastrointestinal tract. Tetracyclines should be administered before meals to prevent formation of such chelates. Antacids that contain cations of calcium, magnesium, or aluminum should not be administered simultaneously with tetracyclines. These cations do not affect the absorption of the other drugs listed in the question.

112. MICROBIOLOGY: ANSWER: B

(Balows, 5/e, pp 985-996) Ebola and Marburg belong to the Filoviridae family. Infections with these two viruses carry a high mortality and have no effective treatment. If either of these agents is suspected, barrier protection and isolation of the patient are necessities. Culture of the virus may be attempted only at a biosafety level 4 laboratory, which can provide maximum containment.

113. PHYSIOLOGY: ANSWER: A

(Ganong, 14/e, p 616. Guyton, 8/e, p 324. Rose, 3/e, pp 102-103. West, 12/e, pp 443, 457-458, and 472-473) Most urea is synthesized in the liver. Its excretion is dependent on its concentration in plasma and the glomerular filtration rate (GFR) in the kidney. Approximately 50 to 60 percent of filtered plasma urea is passively reabsorbed in the proximal tubule at normal GFR. In renal insufficiency, in which GFR is decreased, filtrate remains in the tubules longer and more of the filtered urea is reabsorbed, resulting in an increase in its plasma concentration. Urea is also passively reabsorbed in distal tubules and collecting ducts and recirculates back into tubular urine by means of the countercurrent mechanism.

114. BIOCHEMISTRY: ANSWER: B

(Stryer, 3/e, p 386) Dihydroxyacetone phosphate (DHAP) in the glycolytic pathway is readily reduced to α-glycerol phosphate, which is the source of glycerol in the esterification of fatty acids. Glycerol kinase exists, but it is not the major source of α-glycerol phosphate. Insulin is elevated in the fed state, and thus would not be expected to inhibit the storage of fat.

115. BIOCHEMISTRY: ANSWER: D

(Stryer, 3/e, p 619) Purines are converted via xanthine into uric acid in humans and then excreted. Other species further metabolize uric acid into allantoic acid. Orotic acid is a precursor for the pyrimidines, and urea is the main excreted form of nitrogen.

116. PHYSIOLOGY: ANSWER: B

(Ganong, 14/e, pp 418-420. Guyton, 8/e, pp 702-703) The rate of gastric emptying is regulated by stimuli that originate in both the stomach and the duodenum. The two organs are coordinated to prevent movement of gastric contents at a rate faster than they can be processed by the small bowel. Within the duodenum, distention, increased concentration of hydrogen ion, and inappropriate osmolarity all will retard gastric emptying. Gastrin enhances gastric motility and promotes gastric emptying. Carbohydrate is rapidly hydrolyzed and absorbed and has little effect on the rate of gastric emptying.

117. GENETICS: ANSWER: B

(Gelehrter, pp 57-65. Thompson, 5/e, pp 349-363) Many common disorders tend to run in families but are not single-gene or chromosomal disorders. These disorders are multifactorial traits caused by multiple genetic and environmental factors. The term *polygenic* may be used to refer to this genetic component.

118. BIOCHEMISTRY: ANSWER: B

(Stryer, 3/e, p 268) (A) is biotin; (C) is pantothenic acid; (D) is nicotinic acid or niacin; and (E) is riboflavin. Biotin is required for CO_2 fixation; pantothenic acid is a precursor for coenzyme A; nicotinic acid or niacin is a precursor for NADP; and riboflavin is a precursor for flavin mononucleotide. Ascorbic acid is required for the conversion of proline to hydroxyproline.

119. MICROBIOLOGY: ANSWER: B

(Jawetz, 19/e, pp 434-438) Cytomegalovirus can cause infectious mononucleosis-like disease. However, heterophil antibodies are usually absent. This disease can either occur spontaneously or after blood transfusion.

120. PHYSIOLOGY: ANSWER: B

(Berne, 2/e, pp 60-65. Ganong, 14/e, pp 70-73 and 76-78. Guyton, 8/e, p 489) Release of neurotransmitter from presynaptic cells is accompanied by an influx of calcium from the extracellular fluid. The neurotransmitter then interacts with receptors on the plasma membrane of the postsynaptic cell. In terminals under presynaptic inhibition, the influx of Ca^{2+} is diminished via activation of specific presynaptic receptors. The reduced Ca^{2+} flux causes a reduction in the amount of neurotransmitter released in response to arriving action potentials.

121. PATHOLOGY: ANSWER: A

(Robbins, 4/e, pp 1187-1191) The spectrum of benign breast disease includes fibrocystic disease, which is probably a misnomer; adenosis, both sclerosing and microglandular; intraductal papillomas and papillomatosis; apocrine metaplasia; fibrous stromal hyperplasia; and hyperplasia of the epithelial cells lining the ducts and ductules of the breasts. At one time or another each of the above was considered to be a forerunner of carcinoma; however, with extensive studies in the literature, none of these has been shown to necessarily correlate with a greater risk of developing carcinoma with the exception of epithelial hyperplasia. With any of the features, but especially epithelial hyperplasia, adding a positive family history of breast cancer in a sibling, mother, or maternal aunt markedly increases the risk for developing carcinoma of the breast in the given patient. Owing to the advances and technology of xeromammography, there has been an increased interest in calcifications, which are markers for carcinoma of the breast. These calcifications, however, do not necessarily occur within the cancerous ducts themselves and can be found frequently in either adenosis adjacent to the carcinoma or even in normal breast lobules in the region. Stipple calcification as seen by xeromammography is regarded as an indication for a biopsy of the region by some workers.

122. PATHOLOGY: ANSWER: D

(Robbins, 4/e, pp 45-46) Leukocyte function antigen 1 (LFA-1) is one of the group of adhesive molecules on leukocytes. Other members of this group, all glycoproteins and all having identical β chains, are MO-1 and P-150. A genetic deficiency of the β-chain adhesive proteins on leukocytes (leukocyte adhesion deficiency) results in impaired leukocyte adhesion and recurrent bacterial infections. Endothelial leukocyte adhesion molecule (ELAM-1) is one of the adhesive surface proteins on endothelial cells. Complement fragments (C5a) and leukotriene B4 are mediators that stimulate an increase in adhesive molecules on *leukocytes*. Conversely, interleukin 1 (IL-1) stimulates leukocytic adhesion by action on adhesive surface proteins on *endothelial* cells. Tumor necrosis factor (TNF) stimulates adhesive molecules on *both* leukocytes and endothelial cells.

123. ANATOMY: ANSWER: B

(Hollinshead, 4/e, pp 295-297) Intervertebral disks are strongly reinforced ventrally and laterally by the anterior longitudinal ligaments. The posterior longitudinal ligament, while it is denticulate and attenuated laterally, reinforces the posterior aspect of the intervertebral disk. Because the posterolateral region of the disk is supported least by ligamentous structures, a nucleus pulposus that is herniated through the annulus fibrosus of the intervertebral disk will take the line of least resistance and move posterolaterally into the intervertebral canal. In so doing, the herniation is apt to impinge on a spinal nerve of the next lower vertebral level.

124. BIOCHEMISTRY: ANSWER: C

(Stryer, 3/e, p 412) Phosphorylation sites in the electron transport chain are those places where the free energy of electron transfer is sufficient to form a molecule of ATP from ADP and P_i. These sites are between NAD and flavoprotein, between coenzyme Q and cytochrome *c* (not *b*), and between cytochrome oxidase and oxygen.

125. PHYSIOLOGY: ANSWER: D

126. PHYSIOLOGY: ANSWER: B

(Ganong, 14/e, pp 556-557. Guyton, 8/e, pp 407-409. West, 12/e, pp 575-576) In the diagram accompanying the question, zones I through IV represent gas from different topographical areas of the lung. Zone I represents dead space, zone II is a mixture of dead space and alveolar gas, and zone III represents pure alveolar gas. At point D, the tracer content of expired gas increases. That point represents the closing volume, which is the lung volume at which airways in the lower parts of the lung close off because the transmural pressure gradient is lower owing to the effects of gravity. The gas in apices (zone IV) is richer in the tracer gas because the alveoli in that portion of the lung receive more gas during the early part of inspiration. The closing volume represents approximately 10 percent of vital capacity and increases with age, as does residual volume.

127. BEHAVIORAL SCIENCE: ANSWER: D

(Lerner, 1984. pp 195-207) Motor vehicle accidents are the major cause of death among adolescents and young adults. Physicians consider them to be of epidemic proportion among adolescents—in whom they account for 36 percent of all deaths. Homicide is the second leading cause of death with about 25 percent of homicides occurring in the 15-to-24-year-old group. Suicide is the third leading cause of death and is increasing for this age group, while it is decreasing for the general population. Infections and illicit drug abuse account for a relatively small number of deaths in this age group.

128. BIOCHEMISTRY: ANSWER: D

(Stryer, 3/e, pp 369, 434) Transketolase contains a tightly bound molecule of thiamine pyrophosphate, in contrast with transaldolase which forms a Schiff base via a lysine residue in the course of its reaction. Thiamine deficiency thus compromises transketolase activity.

129. PHARMACOLOGY: ANSWER: B

(DiPalma, 3/e, p 623) Sulfonamides should not be used in neonates, especially premature infants, since the drug competes with bilirubin for serum albumin binding. This results in increased levels of free bilirubin causing kernicterus. Kernicterus is characterized by jaundice, an absent startle reflex, abnormal eye movements, muscle spasam causing the body to bend with convexity forward (opisthotonus). These children also often have a high-pitched cry. Pregnant women at term also should not receive sulfonamides because of this drug's ability to cross the placenta and enter the fetus in concentrations sufficient to produce toxic effects.

130. MICROBIOLOGY: ANSWER: B

(Davis, 4/e, pp 927-928) Parvovirus B19 is the causative agent of erythema infectiosum (fifth disease). It is associated with transient aplastic crisis in persons with hereditary hemolytic anemia. In adults it is also associated with polyarthralgia.

131. BIOCHEMISTRY: ANSWER: B

(Stryer, 3/e, pp 503, 634) Leucine, a purely ketogenic amino acid, and palmitate are degraded to acetyl coenzyme A, which is ultimately converted to CO_2 and can never be used to make glucose in animals. Pyruvate, via pyruvate carboxylase and phosphoenolpyruvate carboxykinase, can be used to make glucose in gluconeogenesis.

132. MICROBIOLOGY: ANSWER: B

(Jawetz, 19/e, p 578) If the patient was infected 2 weeks prior, there is little likelihood of a detectable antibody response. Therefore, either enzyme immunoassay or Western blotting for antibody could be negative in spite of infection. All three of the other tests determine the presence of virus and might be appropriate. However, culture generally costs around $300 and is only 80 percent sensitive. Polymerase chain reaction is a very sensitive test if done correctly, but it costs about $150. The antigen test is the best, most cost-efficient test available.

133. PHYSIOLOGY: ANSWER: C

(Ganong, 15/e, pp 545-546. West, 12/e, pp 128-131) Movement of substances across the capillary wall can occur by diffusion, filtration, or intracellular vesicular transport. Gases such as CO_2 and O_2, which are lipid-soluble, and small, water-soluble substances (e.g., glucose and urea) diffuse readily through the capillary wall, and their concentrations within capillary blood decrease rapidly with distance along the capillary because they equilibrate rapidly. Transcapillary movement of such substances will increase if their transcapillary concentration gradient increases or if capillary flow is increased. Transcapillary movement of larger molecules is dependent on filtration through the capillary pores and the rate of transcapillary movement is inversely related to their molecular size. Such substances do not reach equilibrium across the capillary, and their concentration decreases linearly along the length of the capillary.

134. BIOCHEMISTRY: ANSWER: B

(Stryer, 3/e, pp 374-375) Amino acid activation is the first step in protein synthesis, which proceeds from the *N*-terminal end to the *C*-terminal end, with the amino group of each new amino acid reacting with the carboxyl group of its neighbor. It thus makes sense to activate the new amino acid at its carboxyl end, leaving its amino group free to react. Remember that the formation of the peptide bond requires no discreet energy input. Amino acids do not form coenzyme A derivatives.

135. PATHOLOGY: ANSWER: C

(Robbins, 4/e, pp 58-60, 70-71, 168-169, 183) Two of the major systemic effects of inflammation (acute-phase reactions) are fever and leukocytosis. Both of these reactions are largely under the control of the cytokines IL-1 or α-TNF (cachectin) or both. These two cytokines have many similar functions, including induction of fever, release of ACTH, leukocytosis, and other systemic acute-phase responses. IL-1 initiates fever by inducing synthesis of prostaglandin E_2 (PGE_2) in the anterior hypothalamus, followed by transmission via the posterior hypothalamus, vasomotor center, and sympathetic nerves to cause skin vasoconstriction. Leukocytosis occurs initially because of rapid release of cells from the postmitotic reserve pool of the bone marrow, which is caused by IL-1 and α-TNF and associated with a "shift to the left" of immature cells. β-TNF, another cytokine, has no major role in acute-phase reactions. It takes part in T-cell cytotoxicity and its release may injure cell membranes directly in T-cell-mediated lysis.

136. PHARMACOLOGY: ANSWER: D

(DiPalma, 3/e, pp 611-612) Streptomycin and other aminoglycosides can elicit toxic reactions involving both the vestibular and auditory branches of the eighth cranial nerve. Such reactions are likely to occur in about 75 percent of patients who receive long-term, high-dosage aminoglycoside therapy. Therefore, patients receiving an aminoglycoside should be monitored frequently for any hearing impairment owing to the irreversible deafness that may result from its prolonged use. None of the other agents listed in the question adversely affect the eighth cranial nerve function.

137. ANATOMY: ANSWER: A

(Alberts, 2/e, p 458. Junqueira, 6/e, pp 37-38. Kelly, 18/e, pp 42-47) The Golgi apparatus is usually juxtanuclear and has a typical structure composed of small vesicles in close proximity to the endoplasmic reticulum, lamellar cisternae, and large vacuoles in which secretory material is concentrated. Studies have demonstrated that the Golgi complex is responsible for packaging the products of the endoplasmic reticulum and that it is perhaps also responsible for the posttranslational processing (e.g., glycosylation, sulfation, phosphorylation, amidation) of these products.

138. PHYSIOLOGY: ANSWER: B

(Ganong, 15/e, pp 580-582. West, 12/e, pp 903-904) The ductus arteriosus connects the aorta and pulmonary artery and functions as a physiologic right-to-left shunt during fetal life because the lungs are collapsed and pulmonary vascular resistance is higher than systemic resistance. At birth, the lungs expand and pulmonary vascular resistance and pulmonary artery pressure fall drastically, while systemic vascular resistance and pressure rise owing to removal of the low-resistance placental circulation. Blood flow through the ductus is then reversed. Within a few hours the elevated P_{O_2} in the aortic blood passing through the ductus causes it to constrict and finally to close completely within a few weeks. Since the patency of the ductus is maintained in part by prostacyclin, inhibitors of prostaglandin synthesis such as indomethacin have been used to induce ductus closure and avoid surgery.

139. BEHAVIORAL SCIENCE: ANSWER: A

(Bell, 6/e, pp 397-398) Poverty is the most important social variable in infant mortality. It is linked to an almost 50 percent greater risk of both neonatal and postnatal death. Other social variables—such as the educational level, occupation, or income of either the mother or father, or whether the parents are married—can influence the risk of infant mortality, but none is as powerful as poverty. Poverty is also a major contributor to adult morbidity and mortality.

140. BIOCHEMISTRY: ANSWER: D

(Stryer, 3/e, p 157) DPG lowers the affinity of hemoglobin for oxygen, facilitating the delivery of oxygen to tissues under the condition of low oxygen concentration. Only curve 4 illustrates this; curves 1 and 2 illustrate an increase in affinity.

141. PATHOLOGY: ANSWER: D

(Robbins, 4/e, pp 1158-1164) Malignant ovarian tumors, in the main, arise from the surface epithelium, which takes its origin from cells of the müllerian system. These cells are also referred to as *surface* or *coelomic epithelium*. The müllerian system has the ability to form lining cells of the fallopian tubes, endometrium, and endocervical gland epithelium. Hence, many malignant tumors that take their origin from the surface coelomic epithelium of the ovary resemble these structures. Examples include the borderline serous tumor, serous cystadenocarcinoma, serous cystadenofibrocarcinoma, borderline mucinous tumor, mucinous cystadenocarcinoma, endometrial carcinoma, undifferentiated carcinoma, malignant Brenner tumor, and clear cell adenocarcinoma. Stem cells from the urogenital ridge can give rise to any genitourinary structure. Hilar cells represent small clusters and cords of androgen-producing cells. These presumably give rise to sex cord and Sertoli-Leydig cell tumors. Stromal cell tumors include granulosa cell tumors, theca cell tumors, thecomas, and fibromas. Germ cell tumors give rise to malignant and benign teratomas, including the cystic form (dermoid cyst), dysgerminoma, endodermal sinus tumor, choriocarcinoma, and mixed germ cell tumors.

142. BIOCHEMISTRY: ANSWER: B

(Stryer, 3/e, pp 800-805) Glucose will be preferentially used because in its presence, the lac operon will be turned off; the CAP protein, in the absence of sufficient cAMP to inactivate it, blocks the transcription of genes whose products are required for lactose metabolism.

143. MICROBIOLOGY: ANSWER: E

(Jawetz, 19/e, pp 531-535) Rubella virus does not produce cytopathic effects (CPEs) in tissue-culture cells. Moreover, rubella-infected cells challenged with a picornavirus are resistant to subsequent infection and thus would not exhibit CPEs. Mouse kidney cells infected only with picornavirus would show CPEs.

144. PHYSIOLOGY: ANSWER: D

(Ganong, 15/e, pp 611-614) The pulmonary circulation is a low-pressure system compared with the systemic circulation. Because of this low pressure and the hydrostatic pressure gradient from the top, or apex, of the lung to the bottom, or base, of the lung, the apex of the lung is not as well perfused as the base of the lung. During vigorous exercise there is a large (up to sixfold) increase in cardiac output. The increased flow through the systemic circulation is equal to the increase in flow through pulmonary circulation. Total lung flow increases and flow at the base of the lung is still greater than flow at the apex. However, the flow at the apex, since it was originally low, may increase by up to 800 percent, whereas flow in the base of the lung only increases by up to about 300 percent. The pulmonary vessels are very compliant. The increased flow causes recruitment of previously closed capillaries and dilation of pulmonary arterioles and capillaries. Because of this, pulmonary artery pressure normally does not increase or increases by only a small amount, and it rarely increases more than twofold.

145. PHYSIOLOGY: ANSWER: C

(Ganong, 15/e, pp 585-588) Blood flow to skeletal muscle increases by a factor of 15 to 20 during exercise because the exercising skeletal muscles release metabolites that dilate the arterioles within the muscles. Circulating catecholamines (norepinephrine and epinephrine) act to constrict blood vessels, not dilate them. In addition cardiac output increases and blood is shunted away from nonexercising muscles and other organs, such as the kidney and gastrointestinal tract, that can survive for the duration of exercise with a reduced blood flow. The massive dilation of blood vessels in exercising muscles is greater than the catecholamine-induced constriction in other organs and, as a result, total peripheral resistance decreases. Parasympathetic stimulation to the heart decreases during exercise, which allows the heart rate to increase. The veins and venules are constricted by sympathetic nerve discharge and circulating catecholamines. The increased tone within these muscles increases central venous pressure, which results in an increased preload and increased stroke volume.

146. MICROBIOLOGY: ANSWER: B

(Balows, pp 270-272) Complement fixation and the indirect fluorescent antibody test are both useful in diagnosing acute infections with varicella-zoster virus (VZV); however, they do not have the sensitivity to determine immune status. The direct fluorescent antibody test is a test for antigen, not antibody. Enzyme immunoassay is not a procedure available for VZV. FAMA is usually the best assay for determination of immune status in VZV. Also, neutralization and anticomplement immunofluorescent tests may be used.

147. BIOCHEMISTRY: ANSWER: C

(Stryer, 3/e, p 428) The hexose monophosphate shunt is a source of NADPH and the means by which ribose is ordinarily made from glucose. The NADPH is generated by the action of glucose-6-phosphate dehydrogenase.

148. PHYSIOLOGY: ANSWER: D

(Ganong, 15/e, pp 570-578, 654-655) The kidney receives approximately 20 percent of the cardiac output while consuming a much smaller portion of the body's oxygen consumption. The high renal blood flow is related to the kidney's role in regulating the composition of the extracellular fluid. The skin is another organ in which the relationship between percentage of blood flow and percentage of oxygen consumption does not hold. Blood flow to the skin is primarily related to its role in temperature regulation. Of course in all organs, if oxygen consumption increases, blood flow will increase.

149. HISTOLOGY: ANSWER: B

(Alberts, 2/e, pp 1018-1019. Coe, pp 803-804. Cotran, 4/e, pp 739-743) The patient is suffering from multiple myeloma. In this disease, there are abnormal changes in the bone marrow indicative of altered plasma cell activity and anemia (hemoglobin data and increasing fatigue). These plasma cells produce elevated levels of interleukin 1 (IL-1), which functions as an osteoclast activation factor. The increased IL-1 stimulates osteoclastic activity and results in elevated serum calcium (12.3 mg/dL). The depletion of bone calcium results in lytic lesions of the skull and pelvis as well as the presence of the compression fracture of the spine. The Bence Jones protein represents free immunoglobulin light chains, which occur in the urine of patients with multiple myeloma.

150. PHYSIOLOGY: ANSWER: C

(Ganong, 15/e, pp 526-527) Cardiac output can be measured by using the Fick principle, which asserts that the rate of uptake of a substance by the body (e.g., O_2 consumption in milliliters per minute) is equal to the difference between its concentrations (milliliters per liter of blood) in arterial and venous blood multiplied by the rate of blood flow (cardiac output). This principle is restricted to situations in which arterial blood is the only source of the substance measured. If oxygen consumption by the body at steady state is measured over a period of time and the difference in arterial O_2 and venous O_2 measured by sampling arterial blood and *pulmonary* arterial blood (which is fully mixed venous blood), cardiac output is obtained from the expression

$$\text{cardiac output} = \frac{O_2 \text{ consumption (ml / min)}}{\left[\left(A_{O_2}\right) - \left(V_{O_2}\right)\right] (\text{ml / L})}$$

Substituting the values for the man presented in the question,

$$\text{cardiac output} = \frac{700}{210 - 140} = 10 \text{ L / min}$$

151. PHARMACOLOGY: ANSWER: B

(DiPalma, 3/e, pp 255-256) Central dopamine receptors are divided into D_1 and D_2 receptors. Antipsychotic activity is better correlated to blockade of D_2 receptors. Haloperidol, a potent antipsychotic, selectively antagonizes at D_2 receptors. Phenothiazine derivatives, such as fluphenazine and promethazine, are not selective for D_2 receptors. Bromocriptine, a selective D_2 agonist, is useful in treatment of parkinsonism and hyperprolactinemia. It produces less adverse reactions than nonselective dopamine receptor blockers.

152. MICROBIOLOGY: ANSWER: B

(Davis, 4/e, p 682) If this patient lived in an endemic area for Lyme disease, the clinical presentation would be sufficient for diagnosis. However, laboratory confirmation would be facilitated by detection of specific antibody. Three to four weeks after a tick bite is still early for an increase in IgG antibodies either measured by ELISA or Western blot. IgM response may also be delayed, but appearance of specific IgM antibodies at 4 weeks after a tick bite is not uncommon. A more sensitive technique, Western blotting, would likely be positive with a band at P 41, the flagellar antigen of *Borrelia burgdorferi,* but only if a specific IgM conjugate is used. Culture of the organism is highly unlikely and should not be routinely attempted.

153. PHYSIOLOGY: ANSWER: B

(Berne, 2/e, pp 409-429) The electrocardiogram (ECG) records the conduction of the action potential through the heart. Changes in the rate, rhythm, or conduction pathway are recorded. Changes in the position of the heart in the chest will change the size and shape of the ECG recorded by the various leads. Local areas of ischemia caused by changes in coronary blood flow will cause changes in the action potentials that will be reflected in the shape of the ECG recording. The ECG is unable to detect any changes in the ability of the heart to develop force.

154. BIOCHEMISTRY: ANSWER: B

(Stryer, 3/e, pp 441, 631) Pyruvate carboxylase is activated by acetyl CoA; in the presence of high levels of acetyl CoA, there is no need to make more from precursors, and every reason to conserve the carbons of lactate as glucose. Fructose 1,6-bisphosphatase is activated by citrate (the presence of which in the cytoplasm is a signal of slowed Krebs cycle due to excess acetyl CoA), and inhibited by AMP and fructose 2,6-bisphosphate.

155. BIOCHEMISTRY: ANSWER: D

(Stryer, 3/e, pp 41-42) Blood pH is about 7.4. A weak base that is 91 percent ionized at pH 7.4 must have a pK_a that is one unit higher than this, or 8.4. If you had been told that the drug was a weak acid, the correct answer would have been 6.4.

156. MICROBIOLOGY: ANSWER: B

(Howard, pp 408-409) Patients treated with antibiotics develop diarrhea that, in most cases, is self-limiting. However, in some instances, particularly in those patients treated with ampicillin or clindamycin, a severe, life-threatening pseudomembranous enterocolitis develops. This disease has characteristic histopathology, and membranous plaques can be seen in the colon by endoscopy. Pseudomembranous enterocolitis and antibiotic-associated diarrhea are caused by an anaerobic gram-positive rod, *Clostridium difficile.*

157. ANATOMY: ANSWER: C

(Hollinshead, 4/e, pp 368, 392) The sciatic nerve emerges from the greater sciatic foramen through the infrapiriformis recess. This occurs midway between the posterior superior iliac spine and the ischial tuberosity deep to the gluteus maximus muscle. The sciatic nerve passes into the posterior thigh about midway between the greater trochanter and the ischial tuberosity. It passes deep to the hamstring muscles to enter the popliteal fossa, where it bifurcates into the common peroneal nerve and the tibial nerve. The sciatic nerve may arise as separate tibial and common peroneal nerves, or it may divide into these components at any point within the posterior thigh.

158. MICROBIOLOGY: ANSWER: E

(Howard, p 424) Reports have indicated that the corynebacterium, *Corynebacterium JK (C. jaekium)*, causes significant morbidity in hospitalized patients, especially those with intravenous catheters. *Corynebacterium JK* can be distinguished from other members of the genus by biochemical and cultural tests. Another common characteristic of *JK* is its antimicrobial resistance pattern. *JK* is commonly susceptible only to vancomycin.

159. PATHOLOGY: ANSWER: B

(Anderson, 9/e, pp 1688-1960) Benign cystic teratomas constitute about 10 percent of cystic ovarian tumors. The cysts contain greasy sebaceous material mixed with a variable amount of hair. The cysts' walls contain skin and skin appendages, including sebaceous glands and hair follicles. A variety of other tissues, such as cartilage, bone, tooth, thyroid, respiratory tract epithelium, and intestinal tissue, may be found. The presence of skin and skin appendages gives the tumor its other name, "dermoid" cyst. Dermoid cysts are benign, but in less than 2 percent, one element may become malignant, most frequently the squamous epithelium.

160. CELL BIOLOGY: ANSWER: E

(Junqueira, 7/e, pp 55-57. Moore, pp 83-84. Ross, 2/e, pp 42-43) Females have two X chromosomes, one of maternal and the other of paternal origin. Only one of the X chromosomes is active in the somatic, diploid cells of the female; the other X chromosome remains inactive and is visible in appropriately stained interphase cells as a mass of heterochromatin. Detection of the Barr body (sex chromatin) is an efficient method for the determination of chromosomal sex and abnormalities of X-chromosome number; however, it is not definitive proof of maleness or femaleness. In Turner's syndrome (XO) no Barr bodies would be present. In comparison, "superfemales" (XXX) and persons with Klinefelter's syndrome (XXY) would both possess two inactive X chromosomes, although the sex of these persons would be female and male, respectively, as determined by the presence or absence of the testis-determining Y chromosome. Since only one X chromosome is present in each diploid cell, normal females are mosaic for the X chromosome in that relatively equal clones of X-maternal and X-paternal cells occur.

161. PHARMACOLOGY: ANSWER: B

(DiPalma, 3/e, p 375) Persons with low hepatic *N*-acetyltransferase activity are known as slow acetylators. A major pathway of metabolism of both procainamide, used to treat arrhythmias, and hydralazine, used to prevent hypertension, is *N*-acetylation. Slow acetylators receiving these drugs are more susceptible than normal persons to side effects, since slow acetylators will have higher-than-normal blood levels of these drugs. *N*-Acetylprocainamide, the metabolite of procainamide, is also active and is being tested as an antiarrhythmic agent.

162. BIOCHEMISTRY: ANSWER: A

(Stryer, 3/e, p 992) Aspirin inhibits the production of prostaglandins which are made from polyunsaturated fatty acids by the enzyme cyclooxygenase. Aspirin (acetylsalicylic acid) acetylates a subunit of the enzyme, inactivating it.

163. PATHOLOGY: ANSWER: B

(Robbins, 4/e, pp 1024-1036) Glomerular injury caused by circulating antigen-antibody complexes is a secondary effect from a nonprimary renal source. Numerous clinical examples exist of a serum sickness-like nephritis as a consequence of systemic infection, with classical clinical models such as syphilis, hepatitis B, malaria, and bacterial endocarditis leading to renal disease. Immune complexes to antigens from any of these sources are circulating within the vascular system and become entrapped within the filtration system of the glomerular basement membranes. This can be seen as granular bumpy deposits by immunofluorescence within the basement membranes of the glomeruli. Linear fluorescence, on the other hand, is seen in primary antiglomerular basement membrane disease, wherein antibodies are directed against the glomerular basement membrane itself. Plasma cell interstitial nephritis is seen in immunologic rejection of transplanted kidneys. Nodular glomerulosclerosis is an effect of diabetes mellitus. The presence of red blood cell casts in the urine nearly always indicates that there has been glomerular injury, but is not specific for any given cause. Glomerular basement membrane thickening caused by subepithelial immune deposits is seen in membranous glomerulonephritis. While the morphology of membranous glomerulonephritis is different from that of nephritis caused by circulating antigen-antibody complexes (immune complexes), there are similarities in the pathogenesis in that both disorders may be a consequence of or in association with infections such as hepatitis B, syphilis, and malaria. Other causes for membranous glomerulonephritis include reactions to penicillamine, gold, and certain malignancies such as malignant melanoma.

164. PHYSIOLOGY: ANSWER: E

(West, 12/e, pp 258-259, 307-312) Among the compensatory mechanisms that develop in response to heart failure is an increase in retention of fluid by the kidney. Increased retention of fluid causes the end-diastolic volume of the heart to increase, which, by the Starling mechanism, increases the strength of the heart beat. However, two deleterious effects result from an increase in end-diastolic volume. A larger-than-normal end-diastolic volume causes an increase in end-diastolic pressure, which can lead to pulmonary edema. In addition, the large end-diastolic volume increases the wall stress that must be developed by the heart with each beat, and this increases the myocardial requirement for oxygen. The increase in contractility that results from the administration of a positive inotropic drug such as ouabain will allow the heart to produce the same force at a lower volume and thus eliminate the need for an increase in volume of fluid.

165. ANATOMY: ANSWER: D

(Hollinshead, 4/e, pp 880, 883-884) The palpebral portion of the orbicularis oculi muscle (innervated by the facial nerve) produces the blink. The buccal branch of the facial nerve innervates muscles of facial expression (including the buccinator muscle) between the eye and the mouth, while the buccal branch of the trigeminal nerve is sensory. The levator palpebrae superioris muscle, which elevates the upper eyelid, is innervated by the oculomotor nerve, while the involuntary superior tarsal muscle is supplied by sympathetic nerves.

166. ANATOMY: ANSWER: A

(Hollinshead, 4/e, pp 878-879) The buccinator muscle draws the cheek against the teeth during mastication and with the tongue serves to keep food on the grinding surface of the molars. The buccinator is innervated

by the buccal branch of the facial nerve (CN VII) and paralysis of this muscle results in accumulation of food in the cheek so that mastication becomes awkward and difficult. Because the orbicularis oris muscle is also paralyzed, the patient tends to drool and to have difficulty drinking.

167. ANATOMY: ANSWER: B

(Hollinshead, 4/e, pp 935-936) The auricular branch of the facial nerve (CN VII) innervates a variable portion of the external auditory meatus. These sensory neurons have their cell bodies in the geniculate ganglion and synapse in the brainstem with neurons in the spinal nucleus of cranial nerve V. The geniculate ganglion also contains the cell bodies for the taste fibers from the anterior two-thirds of the tongue, which course in the chorda tympani.

168. MICROBIOLOGY: ANSWER: B

(Mandell, 3/e, pp 1381-1382) While the essential information (i.e., the evidence that the child in question was scratched by her pet cat) is missing, the clinical presentation points to a number of diseases, including cat-scratch disease (CSD). Until recently the etiologic agent of CSD was unknown. Recent evidence indicates that it is a pleomorphic rod-shaped bacterium that has been named *Afipia*. It is best demonstrated in the affected lymph node by a silver impregnation stain. One reason for the repeated failure to isolate this organism is that it apparently loses its cell wall and is very sensitive to the artificial conditions of culture. New information indicates that *Afipia* will grow on artificial media designed for the isolation of *Legionella*.

169. PATHOLOGY: ANSWER: D

(Robbins, 4/e, pp 4-9) Injury to the cell membrane, the loss of cell ATP, and the influx of Ca^{2+} into the mitochondria are thought to be the most critical of multiple cellular events after experimental ischemia in animals. A major detriment following reduction of cell oxygen tension is the cessation or reduction in ATP because of falling oxidative phosphorylation; this occurs early in hypoxia. While increasing anaerobic glycolysis subsequently occurs, the pivotal step is loss of the energy-producing ATP, which leads to cell Na^+ accumulation, K^+ efflux, and Ca^{2+} influx, by reduced effectiveness of the active transport Na^+ pump. Cells greatly enlarge as a consequence of isoosmotic water accumulation. These changes are still reversible if oxygen tension is restored. Continuing hypoxia results in mitochondrial damage (vacuole formation), which is irreversible. Cell death will occur when lysosomes break down and release proteases, RNA and DNAases, and cathepsins.

170. BIOCHEMISTRY: ANSWER: D

(Stryer, 3/e, pp 76, 650) The Watson-Crick model shows a double-stranded structure with hydrogen bonds, not covalent bonds, between A and T and between G and C. Phosphate groups, with their high negative charge, are on the outside of the helix to minimize charge repulsion. The two strands do indeed run in opposite directions, 5′ to 3′ for one strand and 3′ to 5′ for the other.

171. MICROBIOLOGY: ANSWER: A

(Howard, pp 191-192) It has recently been reported that many women may suffer from anterior urethral syndrome (AUS), which is a relatively mild form of urinary tract infection. Unfortunately, the methods used for processing of urine for culture may miss the causative agents of AUS. It has been shown that the most sensitive colony-count breakpoint for these symptomatic patients is not 1×10^5 cfu/mL but 1×10^2 cfu/mL. The causative agent is usually *E. coli,* but *Chlamydia* has been isolated also.

172. BIOCHEMISTRY: ANSWER: C

(Guyton, 8/e, p 74) Muscle cells contain only sufficient ATP to sustain contraction for about one second. When not contracting, muscle cells use the excess ATP to make and store creatine phosphate which serves as an effective ATP buffer.

173. PATHOLOGY: ANSWER: D

(Braunwald, 11/e, p 1582. Robbins, 4/e, pp 1121-1125) In nearly 75 percent of cases, adenocarcinoma of the prostate arises in the posterior lobe, usually in a subcapsular location. The lateral lobes are the next, much less frequent site. Nodular hyperplasia occurs in the periurethral region. When prostatic cancer is extracapsular or metastatic (commonly osteoblastic metastases to pelvis and lumbar vertebrae), serum tumor markers such as prostatic acid phosphatase (PAP) are detectable by standard assays. Tumor growth may be inhibited by estrogen therapy; it is not estrogen-dependent. Invasion of capsule, blood vessels, and perineural spaces is useful in diagnosis of well-differentiated tumor. Diagnosis may include needle biopsy or fine needle aspiration (80 percent accuracy).

174. HISTOLOGY: ANSWER: E

(Junqueira, 7/e, p 104. Ross, 2/e, p 89. Uitto, pp 9, 109, 111, 114, 126) Ehlers-Danlos disorders include type IV disease in which there are problems in the transcription of type III collagen mRNA or in translation of this message. The result is breakdown of the type III collagen in the intestinal and aortic walls, which is responsible for the elasticity of these organs. Hyperextensible skin occurs in Ehlers-Danlos type VI disorder in which problems with the hydroxylation of the amino acid lysine and subsequent cross-linking result in enhanced elasticity. Type VII Ehlers-Danlos disorder involves a specific deficiency in an amino terminal procollagen peptidase. This results from a genetic mutation that alters the propeptide sequence in such a way that the molecular orientation and cross-linking are adversely affected. Increased degradation of proteoglycans occurs in osteoarthritis. Type I collagen is found in dentin.

175. MICROBIOLOGY: ANSWER: C

(Balows, 5/e, pp 442-444) The symptoms of Legionnaires' disease are similar to those of mycoplasmal pneumonia and influenza. Affected persons are moderately febrile, complain of pleuritic chest pain, and have a dry cough. Unlike *Klebsiella* and *Staphylococcus*, *Legionella pneumophila* exhibits fastidious growth requirements. Charcoal yeast extract agar either with or without antibiotics is the preferred isolation medium. While sputum may not be the specimen of choice for *Legionella*, the discovery of small gram-negative rods by direct fluorescent antibody (FA) technique should certainly heighten suspicion of the disease. *L. pneumophila* is a facultative intracellular pathogen and enters macrophages without activating their oxidizing capabilities. The organisms bind to macrophage C receptors, which promote engulfment.

176. PHARMACOLOGY: ANSWER: D

(DiPalma, 3/e, pp 248-250) Cocaine has local anesthetic properties; it can block the initiation or conduction of a nerve impulse. It is biotransformed by plasma esterases to inactive products. In addition, cocaine blocks the reuptake of norepinephrine and epinephrine and, thus potentiates the effects of injected epinephrine. This action also produces central nervous system stimulant effects including euphoria, excitement, and restless-ness. Peripherally, cocaine produces sympathomimetic effects including tachycardia and vasoconstriction. Death from acute overdose can be from respiratory depression or cardiac failure.

177. MICROBIOLOGY: ANSWER: D

(Jawetz, 19/e, p 596) A probe with the characteristics stated in the question is most certainly an RNA probe. Ribosomal RNA is single-stranded; hence no digestion of double-stranded DNA is required. There is much

more ribosomal RNA than DNA, mRNA, or tRNA. Cross-reaction with other bacteria is not surprising, as ribosomal RNA may have some similar sequences. Most probes marketed are DNA probes. Present tests are moderately rapid (2 to 3 h). Probe technology is expected to produce much more rapid (less than 1 h) tests in the near future.

178. MICROBIOLOGY: ANSWER: A
179. MICROBIOLOGY: ANSWER: C
(Jawetz, 19/e, pp 191-192) There have been a number of outbreaks of food poisoning caused by *Listeria monocytogenes. Listeria* is a common inhabitant of farm animals and can be readily isolated from silage, hay, and barnyard soil. Humans at the extremes of age are most susceptible to *Listeria* infection, and food has been implicated as a vehicle. In the outbreak in Maine, it is likely that the cabbage used for the coleslaw was fertilized with animal droppings and not properly washed prior to consumption. Major *Listeria* outbreaks associated with cheese have been seen in the United States and most likely have originated from contaminated milk. Epidemiologic investigation often will provide data on attack rates in such outbreaks. The eventual solution of the problem always lies in a combination of epidemiologic, microbiologic, and clinical information. For example, in the Maine case, it should not be assumed that the eclairs were the culprit based on the fact that everyone ate them.

180. BIOCHEMISTRY: ANSWER: C
(Stryer, 3/e, p 316) If two reactions are coupled and both proceed in the direction written, their free energies may be summed: +1.5 kcal + (−4.5 kcal) = −3.0 kcal. The overall reaction is exothermic and can proceed spontaneously.

181. PHARMACOLOGY: ANSWER: C
(AMA Drug Evaluations, 7/e, pp 239-242) Mild analgesics such as aspirin and acetaminophen, sedatives, and antianxiety agents may provide nonspecific symptomatic relief for mild migraine headaches. Ergotamine is the most effective specific relief because of its vasoconstricting activity. Ergotamine is often combined with caffeine, which increases its oral absorption and also has cerebral vasoconstricting activity. Propranolol, a β-adrenergic receptor blocker, and methysergide, a serotonergic receptor blocker, are useful in prophylactic therapy. Amitriptyline, a tricyclic antidepressant, and clonidine, an antihypertensive agent, have been used with some success in prophylactic therapy also.

182. PATHOLOGY: ANSWER: D
(Robbins, 4/e, pp 1282-1283) There appear to be strong immune factors that presumably account for some well-documented remissions, lengthy survival after distant metastasis, and rapid growth in renal transplant patients. However, most patients with this form of cancer pursue a course characterized by eventual distant and visceral metastasis, especially if the histologic type is either nodular or superficial spreading. The subtype called lentigo maligna melanoma, found in the sun-exposed skin of elderly patients, generally has a much more favorable outlook. The most important predictors of outcome are the level of penetration into the subepidermis and reticular dermis (Clark levels I through V: I, *in situ*; V, invasion of subcutaneous fat) and the actual depth of invasion, measured in millimeters with an ocular micrometer (Breslow depth). The survival at 5 years is 90 percent if the tumor is Clark I or II and 0.76 mm or less in depth, but survival falls to 40 to 48 percent if the tumor is level III or IV and greater than 1.9 mm in depth. While some melanoma cells may show cytologic pleomorphism, many aggressive melanomas exhibit uniformity and blandness. Recent work has shown that melanomas arising in the region of the shoulder, upper trunk, and back in men behave in an aggressive fashion.

183. BIOCHEMISTRY: ANSWER: D

(Stryer, 3/e, p 630) Lactate is formed from pyruvate under anaerobic conditions in order to reoxidize NADH to NAD⁺. When oxygen is supplied to cells, pyruvate is the end product of glycolysis.

184. PATHOLOGY: ANSWER: D

(Robbins, 4/e, pp 633-637) The successful outcome of treatment of infective endocarditis is directly dependent on early recognition and diagnosis, since with time the infective organisms (such as yeast, bacteria, rickettsiae) tend to be covered with fibrin and platelets, thereby preventing access of antibiotics to the organisms. In addition, delayed treatment allows time for local valvular destruction. Acute endocarditis (AC) tends to develop within days on previously normal valves (60 percent of cases), whereas subacute endocarditis (SEC) takes more time to develop, may be clinically silent, and may be manifested only by the patient complaining of "not being up to par." The organism causing AC tends to be pathogenic (e.g., *Staphylococcus aureus* or gonococcus); the organism causing SEC tends to be relatively innocuous (e.g., microaerophilic streptococci, *Streptococcus viridans*, and even diphtheroids). Bacterial embolization from the valves to other organs occurs mainly in AC and is much less common in SEC. Despite small differences, such as the smaller vegetations occurring in SEC than in AC, it is not usually possible to distinguish SEC vegetations from AC vegetations through structural criteria only.

185. ANATOMY: ANSWER: D

(Langman, 6/e, pp 245-246. Moore, 4/e, pp 224-225) The dorsal pancreatic bud is the source of the body, tail, and isthmus of the pancreas, as well as of the upper half of the head. It also gives rise to the accessory pancreatic duct (of Santorini). The ventral pancreatic bud, which migrates to the right with the lower end of the common bile duct and fuses with the dorsal primordium, forms the uncinate process and the lower half of the head of the pancreas. The *proximal* portion of the *ventral* pancreatic duct anastomoses with the *distal* portion of the *dorsal* pancreatic duct to form the primary pancreatic duct (of Wirsung). The primary duct, in conjunction with the bile duct, most commonly opens into the duodenal papilla (of Vater).

186. BEHAVIORAL SCIENCE: ANSWER: D

(Mussen, 6/e, pp 64-65) Fetal alcohol syndrome can occur as the result of chronic heavy drinking by a pregnant woman. Symptoms may include severe retardation of intrauterine growth, premature birth, microcephaly, and other deformities such as congenital eye and ear problems, heart defects, extra fingers and toes, and patterns of disturbed sleep. Most recent research suggests a 10 percent risk of this syndrome if the pregnant woman drinks as little as 2 to 4 ounces of hard liquor daily. It is estimated that 6000 infants a year suffer from the effects of fetal alcohol syndrome.

187. PATHOLOGY: ANSWER: D

(Fitzpatrick, pp 46-55. Robbins, 4/e, pp 1300-1301) All the features mentioned plus edema with clubbing of dermal papillae and dilatation of straight capillaries in the dermal papillae are, taken together, pathognomonic, histologic characteristics of psoriasis. The diagnosis of a lesion as "psoriasiform dermatitis" indicates that some but not all of the six characteristics are present. A partial pattern is nondiagnostic and should not be taken to mean psoriasis, since it is a nonspecific histologic pattern that may be found in other conditions, including exfoliative dermatitis, seborrheic dermatitis, chronic contact dermatitis, and neurodermatitis.

188. BIOCHEMISTRY: ANSWER: C

(Stryer, 3/e, pp 7, 10) The binding of a peptide hormone to its receptor is a freely reversible event. You would therefore not expect the formation of covalent bonds, but all of the other responses are plausible.

189. BEHAVIORAL SCIENCE: ANSWER: B

(Conger, 4/e, pp 194-202) Certain children are singled out for abuse by their parents; these children frequently were born prematurely and were slower to develop than were their siblings. Child abuse is committed most often on children below the age of 3. A principal finding in studies of the personality of parents who abuse their children is that such parents usually were abused by their own parents. Mothers abuse their children more often than do fathers, probably because mothers have more contact with their children.

190. PHYSIOLOGY: ANSWER: A

(Berne, 2/e, pp 774-775 and 969-973. Guyton, 8/e, pp 845-846. West, 12/e, pp 481-483) The synthesis and secretion of aldosterone are dependent primarily upon the renin-angiotensin system. Decreased sodium concentration and increased potassium concentration stimulate aldosterone secretion in the absence of changes in plasma volume. The role of adrenocorticotropic hormone (ACTH) in controlling aldosterone synthesis and secretion is negligible.

191. BEHAVIORAL SCIENCE: ANSWER: C

(Williams, 3/e, pp 144-147) Physicians affiliated with health maintenance organizations have a strong financial incentive to provide medical care to enrolled members in a manner that emphasizes efficiency and economy. Prepayment for maintenance of health dictates minimal expenditures for patient care in order to maximize income. Means of accomplishing this include reducing the time patients spend in the hospital, doing as much diagnosis and therapy as possible on an outpatient basis, detecting disease early, and emphasizing preventive services.

192. BIOCHEMISTRY: ANSWER: A

(Stryer, 3/e, pp 458-462, 637) Glycogen is a storage form of glucose. Thus, to raise blood glucose, it is necessary to inhibit the formation of glycogen (B) and increase its breakdown (C). Another means of raising blood glucose is to convert other substances to glucose via gluconeogenesis, which must not be inhibited (D).

193. PHYSIOLOGY: ANSWER: A

(Ganong, 15/e, pp 574-575. Guyton, 8/e, pp 237-244. West, 12/e, pp 267-272) Coronary blood flow is regulated by both chemical and neural factors. Decreased arterial P_{O_2} and increased P_{CO_2} both increase coronary blood flow even in denervated hearts. Both vagal and β-adrenergic stimulation result in coronary vasodilatation. When systemic blood pressure falls, reflex noradrenergic stimulation causes an increase in coronary blood flow and renal splanchnic and cutaneous vasoconstriction. When β-adrenergic receptors are blocked, noradrenergic stimulation mediated by α-adrenergic receptors results in coronary vasoconstriction.

194. BIOCHEMISTRY: ANSWER: E

(Stryer, 3/e, pp 117-139) Western blotting refers to the use of antibodies to detect the presence of specific proteins; it has nothing to do with nucleic acids. All of the other techniques referred to in the question can be used in the construction of a recombinant DNA molecule.

195. PHARMACOLOGY: ANSWER: D

(DiPalma, 3/e, pp 654-655) More than 15 drugs, including dapsone, primaquine, and sulfonamides, have been associated with hemolysis in persons who are deficient in glucose-6-phosphate dehydrogenase. The mechanism of hemolysis is thought to be related to drug-induced free radicals that denature hemoglobin and lead to oxidation of intrachain disulfide bridges. Hemoglobin is precipitated, and red blood cells are destroyed. With adequate glucose-6-phosphate dehydrogenase, oxidation is prevented by maintaining sufficient levels of reduced cofactors (NADH and NADPH) and (consequently) reduced glutathione. The defect is an example of genetically determined, individual variations in response to certain drugs and is most common in persons who belong to ethnic groups originating in the Mediterranean basin. The adverse reactions of aminoglycosides (e.g., gentamicin) include ototoxicity, nephrotoxicity, vestibular toxicity, and neuromuscular blockade.

196. MICROBIOLOGY: ANSWER: E

(Murray, p 671) Macrophages may be activated by CD4+ T cells, and microbial products such as LPS and muramyl dipeptide may also serve as macrophage activators. Activated macrophages express MHC-II proteins on their surface and serve to present antigen to CD4+ T cells. CD4+ T cells also induce suppressor cell function, B cell function, and natural killer (NK) cell function. CD4+ T cells do not suppress cytotoxic T cell function.

197. MICROBIOLOGY: ANSWER: D

(Davis, 4/e, pp 745-746) Fungi are eukaryotic, whereas bacteria are prokaryotic. Fungi are sensitive to griseofulvin; bacteria are not. Bacterial cell walls contain peptidoglycan, fungi do not. The nuclear material of fungi is membrane-bound; in bacteria it is not.

198. BIOCHEMISTRY: ANSWER: A

(Stryer, 3/e, p 374-375) A H^+ and e^- pair is generated by an oxidative step, such as those marked by letters B, C, D, and E on the diagram. Step A is only a rearrangement, or isomerization.

199. BIOCHEMISTRY: ANSWER: C

(Stryer, 3/e, pp 994-995) Proinsulin is converted to insulin via proteolytic action in the β cells of the pancreas prior to secretion. All of the other statements are correct.

200. BEHAVIORAL SCIENCE: ANSWER: E

(Schuster, 2/e, pp 407-413) Although a child's discovery of death is a private and individual experience, it is closely related to cognitive and affective developmental stages. Cognitively, a child must be able to conceptualize animate versus inanimate objects, comprehend cause-effect relationships, and deal with concrete factors before dealing with the abstract. Affectively sufficient ego-strength, separateness, uniqueness, vulnerability, and coping skills must be developed. The infant must develop a concept of self before being able to comprehend "me—not me."

Between birth and 2 years of age, the infant is aware of separation through object loss and separation anxiety. This separation or loss or deprivation is an early form of experiencing perceived as something synonymous with death. Children 3 and 4 years of age believe that all things think, feel, and experience things as they do. Thus they think that a toy feels pain when it is broken and that it must feel hurt when it is repaired. Also, they consider death as another form of life. At 4 and 5 years of age they believe that death is a cessation of movement, but that the dead person or animal continues to experience feelings. Cross-cultural studies have identified many primitive peoples as conceptualizing death on this level, since many of them place food, drink, and other objects at the grave or in the tomb of the deceased. Our custom of placing flowers and other adornments on graves may, in part, be a remnant of this level of conceptualizing death. Children at this age are also apt to associate death with retaliation or punishment, which may lead them to fear, anger, and aggression toward others. Physicians must be aware of this possible interpretation of the 5- and 6-year-old child as a reaction to death, especially in the child's own family.

At 5 or 6 years of age children are apt to feel that death can be avoided and that they are not responsible for another person's death. They also begin to realize that their own significant others can die, and this can lead to notable uncertainties and insecurities. Children at 7 to 9 years of age begin to accept the inevitability of death for all living things, but believe that somehow it is external to oneself.

By 10 to 12 years of age, death begins to be accepted as a biological finality and understood in relation to natural laws, rather than being perceived as the result of aggression or trauma. Adolescents and even adults sometimes find it difficult to accept the concept of nonexistence, even though they are capable of abstract thought.

201. CELL BIOLOGY: ANSWER: C

(Alberts, 2/e, pp 464-465) In inclusion cell disease, there is an absence or deficiency of *N*-acetylglucosamine phosphotransferase. This results in mis-sorting of lysosomal enzymes to the secretory pathway since the absence of phosphorylation in the *cis*-Golgi prohibits segregation of lysosomal enzymes that normally occurs in the *trans*-Golgi through the action of mannose-6-phosphate receptors. Lysosomal enzymes are secreted into the bloodstream, and undigested substrates build up within the cells. There is normal expression of the genes encoding the hydrolases, but a mis-direction of the intracellular sorting signal for these hydrolytic enzymes.

202. HISTOLOGY: ANSWER: B

(Alberts, 2/e, pp 1003, 1008-1009, 1038-1039. Roitt, 2/e, pp 16.1-16.7. Ross, 2/e, p 310) During a viral infection, both cell-mediated and humoral responses are stimulated. In these responses, macrophages phagocytose virus. Cells that become infected with virus can be killed by CD8+ cytotoxic T cells, which can react to the antigen in the presence of MHC class I molecules. T- and B-cell areas of the spleen and lymph nodes will be involved in the filtration of the blood and lymph, respectively. B-cell differentiation requires the presence of CD4+ helper T cells and an antigen-presenting cell. The antigen-presenting cell will phagocytose the virus and present it to helper T cells in the presence of MHC class II molecules. The B cell also presents antigen in this arrangement. It will form a plasma cell and a memory B cell. Activated T cells also enlarge to form large lymphocytes and subsequently undergo cell proliferation to form T cells and memory T cells.

203. GENETICS: ANSWER: E

(Gelehrter, p 15. Thompson, 5/e, pp 46-47) The "genetic code" uses three base "words," or codons, to specify the 20 different amino acids. Since there are four nucleotides that can be arrayed in 2^4 different combinations (64 total), the code must be degenerate, with more than one codon for a single amino acid. Sixty-one codons code for amino acids; three codons are "stop" codons and result in chain termination. The genetic code is universal with codons coding for the same amino acids in all organisms.

204. MICROBIOLOGY: ANSWER: B
(Balows, pp 235-237) With an acute case of primary infection by Epstein-Barr virus (EBV), such as infectious mononucleosis, IgM and IgG antibodies to VCA should be present. Antibodies to EBNA should be absent as they usually appear 2 to 3 months after onset of illness. Culture is not clinically useful because it (1) requires freshly fractionated cord blood lymphocytes, (2) takes 3 to 4 weeks for completion, and (3) is reactive in the majority of seropositive patients.

205. BIOCHEMISTRY: ANSWER: D
(Stryer, 3/e, pp 610-614) If the glycosidic bond were broken, there is no known chemistry for re-forming it that uses the same ribose moiety. Recall that de novo synthesis of ribonucleotides involves the use of phosphoribosyl pyrophosphate (PRPP). The other statements are true.

206. PHYSIOLOGY: ANSWER: C
(Ganong, 14/e, pp 561-562. Guyton, 8/e, pp 404-405. West, 12/e, pp 536-537) The lung has many metabolic functions, including synthesis of surfactant and prostaglandins and withdrawal (inactivation) of prostaglandins and bradykinin from the circulation. Other metabolic functions include activation of angiotensin I to angiotensin II, release of histamine, and inactivation of serotonin. Prostaglandins E and F both are synthesized and removed from the circulation by the lungs.

207. ANATOMY: ANSWER: E
(Junqueira, 6/e, pp 143-144. Kelly, 18/e, p 203) The two forms of bone distinguishable under low magnification are compact bone and trabecular, or cancellous, bone. Trabecular bone, as shown in the photomicrograph accompanying the question, consists of a three-dimensional lattice of branching spicules or trabeculae lining a system of intercommunicating spaces filled with bone marrow. In the long bones the shaft consists of a thick-walled, hollow cylinder of compact bone; trabecular bone occurs in the marrow cavity, the metaphyses, and the epiphyses. In flat bones the inner and outer plates are composed of compact bone; the central region, the diploe, is trabecular bone.

208. HISTOLOGY: ANSWER: B
(Alberts, 2/e, p 650. Junqueira, 7/e, pp 47, 64, 337-341. Ross, 2/e, p 66. Stevens, p 35. Widnell, p 186) In *immotile cilia syndrome* the outer dynein arms may be absent and microtubular arrangements are abnormal. The result is a failure of normal ciliary action. Chronic bronchial and sinus infections are common occurrences in these patients because the cilia are unable to remove foreign materials from the bronchi and sinuses. Infertility in the male may be due to absence of normal ciliary proteins in the flagella of the spermatozoa. Infertility in the female may be related to problems in movement of the ovum through the oviduct. Stereocilia are immotile; they are more similar to microvilli than cilia and are not involved in the movement of sperm despite their location in the epididymis and vas deferens. Many of the patients diagnosed with immotile cilia syndrome are observed to have a lateral transposition of the major organs of the body (situs inversus). Normal ciliary action may be required for normal positioning of organs during development.

209. BIOCHEMISTRY: ANSWER: A
(Stryer, 3/e, pp 233, 244, 459) Aspartate transcarbamoylase is the classic example of allosteric regulation. Chymotrypsin must be activated by proteolysis from chymotrypsinogen, and the others are activated or inhibited by phosphorylation.

210. PHARMACOLOGY: ANSWER: B

(DiPalma, 3/e, pp 352, 355-356) Although ventricular extrasystole is a common arrhythmia associated with digitalis therapy, all the arrhythmias listed in the question may occur in digitalis intoxication. Ventricular fibrillation is the most common cause of death in patients who become toxic to digitalis. Other consequences of digitalis intoxication include blurred, "frosted," or "white" vision neuralgia (particularly trigeminal); malaise, anorexia, and nausea with vomiting; and (only rarely) rashes, eosinophilia, and gynecomastia. Digitalis decreases AV conduction.

Book B

Correct Answers

Number of items: 210

211. ANATOMY: ANSWER: I
212. ANATOMY: ANSWER: H
213. ANATOMY: ANSWER: E
214. ANATOMY: ANSWER: I
(*Moore, 4/e, pp 170-184*) See table below.

TABLE OF PRINCIPAL BRANCHIAL DERIVATIVES

	Groove	Arch	Pouch
I	Pinna and external auditory meatus	Mandible, maleus, incus, anterior part of tongue, mm. of mastication, tensors tympani and veli palatini mm., mylohyoid m., ant. belly of digastric m., trigeminal nerve	Auditory tube, middle ear cavity
II		Lesser horns of hyoid, styloid process, stapes, m. of facial expression, stapedius m., stylohyoid m., post. belly of digastric m., facial nerve	Tonsillar fossa
III		Gr. horns of hyoid, post part of tongue, stylopharyngeus m., glossopharyngeal nerve	Vallecular recess. thymus gland, inf. parathyroids
IV		Thyroid cartilage, cricothyroid m., sup. laryngeal nerve	Sup. parathyroids
V			Ultimobranchial bodies (parafollicular cells)
VI		Cricoid and arytenoid cartilages, intrinsic laryngeal mm., inf. laryngeal nerve	Laryngeal ventricle

215. PHARMACOLOGY: ANSWER: F

(DiPalma, 3/e, pp 583-591. Katzung, 4/e, pp 553-558) Because of its long duration of action, benzathine penicillin G is given as a single injection of 1.2 million units intramuscularly every 3 or 4 weeks for the treatment of syphilis. This persistence of action reduces the need for repeated injections, costs, and local trauma. Benzathine penicillin G is also administered for group A, beta-hemolytic streptococcal pharyngitis and pyoderma.

216. PHARMACOLOGY: ANSWER: A

(DiPalma, 3/e, pp 583-591. Katzung, 4/e, pp 553-558) Methicillin is a β-lactamase-resistant (penicillinase-resistant) penicillin that is acid-labile but must be administered by the parenteral route. It is effective against nearly all strains of *Staphylococcus aureus*. Methicillin is much more effective against penicillinase-producing strains than is penicillin G.

217. PHARMACOLOGY: ANSWER: C

(DiPalma, 3/e, pp 583-591. Katzung, 4/e, pp 553-558) Piperacillin along with mezlocillin and azlocillin is commonly referred to as an extended-spectrum penicillin because of its broad spectrum of activity. It is particularly effective against *Klebsiella* and *Pseudomonas*. Piperacillin is available as a powder for solubilization and injection.

218. PHYSIOLOGY: ANSWER: M
219. PHYSIOLOGY: ANSWER: H
220. PHYSIOLOGY: ANSWER: C
221. PHYSIOLOGY: ANSWER: L

Hormones can be divided into two major groups on the basis of their mechanisms of action and biochemical properties: (1) the lipid-soluble hormones, which include adrenal (cortisol) and gonadal steroids, the iodothyronines (thyroxine), and vitamin D (cholecalciferol), and (2) the water-soluble peptide (LH), thyrotropin-releasing hormone, insulin and catecholamine (epinephrine) hormones.

The peptide and catecholamine hormones are water-soluble and circulate unbound in the plasma and extracellular fluid. These hormones bind to cell surface receptors in their target tissues and do not require penetration of the plasma membrane for their actions. In some cases, hormone-receptor complexes are internalized and degraded via lysosomal hydrolysis. Following interaction with their receptors, these hormones stimulate production of intracellular mediators, which function as "second messengers" for the "first messenger" (hormone). For many peptide and catecholamine hormones the intracellular messenger is a cyclic nucleotide, i.e., cyclic 3′,5′-adenosine monophosphate (cAMP) or cyclic 3′,5′-guanosine monophosphate (cGMP).

Oxytocin is a posterior pituitary peptide that promotes contraction of the myoepithelial cells surrounding mammary gland ducts and causes expulsion of milk from lobular alveoli. Secretion of oxytocin is promoted by tactile stimulation of the breast by the nursing infant. It can also be elicited by psychic factors alone, such as the anticipation of nursing brought on by hearing the cry of the hungry infant. This anticipatory secretion of oxytocin may be experienced by the mother as a sensation of milk let-down in which milk appears at the nipple and may be forcibly ejected.

Somatostatin released locally affects the secretion of both the A and B cells, an example of paracrine regulation.

Secretin is secreted by cells in the duodenal mucosa and regulates pancreatic exocrine secretion. It is responsible primarily for bicarbonate and fluid secretion.

222. HISTOLOGY: ANSWER: F
223. HISTOLOGY: ANSWER: C
224. HISTOLOGY: ANSWER: A
225. HISTOLOGY: ANSWER: F
226. HISTOLOGY: ANSWER: D
227. HISTOLOGY: ANSWER: F
228. HISTOLOGY: ANSWER: I

(Junqueira, 7/e, pp 107-120. Roitt, 2/e, pp 19.8-19.12. Ross, 2/e, pp 92-102. Stevens, pp 50-51, 71, 74-77) The cells of connective tissue include resident cells and migrating cells. The developmental source of these cells also varies with some cells derived from undifferentiated mesenchymal cells and others from the marrow. These cell types include fibroblasts, macrophages, mast cells, adipocytes, neutrophils, eosinophils, basophils, plasma cells, and lymphocytes. The fibroblast is the most common connective tissue cell. Fibroblasts are responsible for the synthesis of the fiber (collagen, elastic, and reticular) and ground substance (glycoproteins and proteoglycans) constituents of the connective tissue matrix. Macrophages are phagocytic cells that originate in marrow, pass through the bloodstream as monocytes, and ultimately enter tissue. They phagocytose bacteria and viruses, a process initiated by complement and IgG. Macrophages phagocytose antigen and secrete it onto their cell surfaces, where it is presented to other cells, including T and B lymphocytes. In different tissues the phagocytes or macrophages may have different names that reflect an independent discovery by a scientist and therefore an eponym for the cell type (e.g., Kupffer cells in the liver, Langerhans cells in the skin, and Hofbauer cells in the placenta).

Eosinophils are phagocytes that appear to be specific for antigen-antibody complexes. They also secrete histaminase, which degrades histamine released from basophils and mast cells, and arylsulfatases, which break down leukotrienes. By carrying out these functions, eosinophils modulate the mast cells and basophils that respond during allergic reactions and provide a means of negative feedback regulation of allergic responses. They are attracted to a site of inflammation by eosinophil-chemoattractant factors (ECFs) released by mast cells and basophils.

Basophils are derived from a different stem cell than the mast cells but closely resemble mast cells in structure. They are the only blood source of histamine.

Plasma cells are derived from B lymphocytes after initiation of the humoral response by macrophages, which phagocytose and present antigen (antigen-presenting cells) and T-helper cells. They are not blood-borne cells, but they secrete antibodies (immunoglobulins) into the bloodstream. They produce all the immunoglobulins: IgG, IgA, IgM, IgD, and IgE.

229. PHARMACOLOGY: ANSWER: D
(DiPalma, 3/e, pp 362, 371. Katzung, 4/e, pp 173-174) Flecainide is related to local anesthetics and also affects sodium channels, but has little effect on repolarization.

230. PHARMACOLOGY: ANSWER: H
(DiPalma, 3/e, pp 362, 371. Katzung, 4/e, pp 173-174) Verapamil is a calcium channel blocker that affects the resting potential or phase 4 and thus has its greatest effect on pacemaker tissue. It is mainly of use in supraventricular arrhythmias.

231. PHARMACOLOGY: ANSWER: E
(DiPalma, 3/e, pp 362, 371. Katzung, 4/e, pp 173-174) Mexiletine, which is in the same group of local anesthetics as lidocaine, is remarkable because it either does not affect or shortens repolarization. Its action is mainly on depolarized fibers.

232. PATHOLOGY: ANSWER: D
233. PATHOLOGY: ANSWER: F
(*Robbins, 4/e, pp 956, 1397, 1425, 1431, 1445-1446*) In Pick's disease, a presenile dementia, neurofibrillary tangles are also demonstrable, but they are fewer than in either Alzheimer's disease or postencephalitic parkinsonism. Pick's disease is a major cortical degenerative disease associated with marked atrophy of frontal lobes and partial temporal lobe atrophy.

Schwannomas (neurilemmomas) are single, encapsulated tumors of nerve sheaths, usually benign, occurring on peripheral, spinal, or cranial nerves. The acoustic neuroma is one example. Microscopically, Verocay bodies, which are foci of palisaded nuclei, may be found in the more cellular (Antoni A) tissue.

234. MICROBIOLOGY: ANSWER: B
235. MICROBIOLOGY: ANSWER: C
236. MICROBIOLOGY: ANSWER: D
237. MICROBIOLOGY: ANSWER: E
(*Balows, 5/e, pp 222-258, 277-286, 360-383, 454-456, 471-477. Howard, p 439*) Diphtheria, a disease caused by *Corynebacterium diphtheriae*, usually begins as a pharyngitis associated with pseudomembrane formation and lymphadenopathy. Growing organisms lysogenic for a prophage produce a potent exotoxin that is absorbed in mucous membranes and causes remote damage to the liver, kidneys, and heart; the polypeptide toxin inhibits protein synthesis of the host cell. Although *C. diphtheriae* may infect the skin, it rarely invades the bloodstream and never actively invades deep tissue. Diphtheria toxin (DT) kills sensitive cells by blocking protein synthesis. DT is converted to an enzyme that inactivates elongation factor 2 (EF-2), which is responsible for the translocation of polypeptidyl-tRNA from the acceptor to the donor site on the eukaryotic ribosome. The reaction is as follows:

$$NAD + EF\text{-}2 = ADP\text{-}ribosyl - EF\text{-}2 + nicotinamide + H^+$$

Bordetella pertussis and *B. parapertussis* are similar and may be isolated together from a clinical specimen. However, *B. parapertussis* does not produce pertussis toxin. Pertussis toxin, like many bacterial toxins, has two subunits: A and B. Subunit A is an active enzyme and B promotes binding of the toxin to host cells.

Francisella tularensis is a short, gram-negative organism that is markedly pleomorphic; it is nonmotile and cannot form spores. It has a rigid growth requirement for cysteine. Human tularemia usually is acquired from direct contact with tissues of infected rabbits but also can be transmitted by the bites of flies and ticks. *F. tularensis* causes a variety of clinical syndromes, including ulceroglandular, oculoglandular, pneumonic, and typhoidal forms of tularemia.

The pathogenesis of infection with *Escherichia coli* is a complex interrelation of many events and properties. *E. coli* may serve as a model for other members of the Enterobacteriaceae. Some strains of *E. coli* are enteroinvasive (EIEC), some enterotoxic (ETEC), some enterohemorrhagic (EHEC), and others enteropathogenic (EPEC). At the present time, there is little clinical significance in routinely discriminating the various types, with the possible exceptions of the ETEC and the 0157/H7 *E. coli* that are hemorrhagic. *E. coli* 0157/H7 secretes a toxin called "*verotoxin.*" The toxin is very active in a Vero cell line. More correctly, the toxin(s) should be called "*Shiga-like.*"

238. PATHOLOGY: ANSWER: D
239. PATHOLOGY: ANSWER: C
(Robbins, 4/e, pp 643-648) The cardiomyopathies (CMP) may be classified into primary and secondary forms. The primary forms are mainly idiopathic (unknown cause). The causes of secondary CMP are many and range from alcoholism (probably the most common cause in the United States) to metabolic disorders to toxins and poisons. Whereas there are not many gross organ and microscopic anatomic features of CMP, a few rather characteristic hallmarks are well recognized in separating the types. However, extensive clinical, historical, and laboratory data contribute as much if not more to classification of the type of CMP present than does biopsy or even the postmortem heart examination.

Hypertrophic CMP encompasses those cases in which the major gross abnormality is to be found within the interventricular septum, which is usually thicker than the left ventricle. If there is obstruction of the ventricular outflow tract, there will be moderate hypertrophy in the left ventricles as well, but the septum usually remains thicker, yielding an appearance of asymmetric hypertrophy. This form of CMP occurs in families (rarely sporadically) and is thought to be autosomal dominant. Up to one-third of these patients have been known to die sudden cardiac deaths, often under conditions of physical exertion. Histologically, the myofibers interconnect at angles and are hypertrophied.

Endomyocardial fibrosis is a form of restrictive CMP found mainly in young adults and children in Southeast Asia and Africa, where it accounts for a significant number of deaths in these age groups. It differs from endocardial fibroelastosis in the United States in that elastic fibers are not present. Its cause is totally unknown.

240. PHARMACOLOGY: ANSWER: H
(DiPalma, 3/e, pp 96, 121, 224, 255, 267, 301, 309, 539) Figure **H** is isoproterenol, a representative of a large class of sympathomimetic compounds, many of which are catecholamine derivatives that stimulate β-adrenergic receptors in bronchiolar smooth muscle, thus inducing the bronchioles to relax. Isoproterenol and structurally similar compounds (epinephrine, albuterol, terbutaline, metaproterenol, and isoetharine) are important bronchodilators used in the therapy of chronic obstructive pulmonary diseases, such as bronchial asthma.

241. PHARMACOLOGY: ANSWER: C
(DiPalma, 3/e, pp 96, 121, 224, 255, 267, 301, 309, 539) Figure **C** illustrates codeine, a derivative of morphine, which contains the pentacyclic opioid structure. Although principally important for their analgesic activity, many opioids are also useful as antidiarrheals, respiratory depressants, and cough suppressants.

242. PHARMACOLOGY: ANSWER: G
(DiPalma, 3/e, pp 96, 121, 224, 255, 267, 301, 309, 539) This structure is the phenothiazine nucleus, attachments to the nitrogen atom which account principally for the differences in the pharmacokinetics of the various compounds that comprise this group. Examples include the piperazines, such as trifluoperazine; the piperidines, such as thioridazine; and the propylamines, or open-chain compounds, such as chlorpromazine. All of these agents are effective antipsychotic drugs. Amitriptyline is shown in Figure **D**. This compound is a member of a large group of compounds known as *tricyclic antidepressants*. The three-ring structure is similar to that of the phenothiazines (Figure **G**), but the pharmacologic properties are quite different. Other drugs included in this group are imipramine, desipramine, trimipramine, doxepin, and nortriptyline.

243. PHARMACOLOGY: ANSWER: A
(DiPalma, 3/e, pp 414-415, 490-495) Levodopa is converted to dopamine in the peripheral tissues by dopa decarboxylase, which has pyridoxine as a cofactor. Excess of this vitamin will increase this reaction, which is an undesirable effect because dopamine does not cross the blood-brain barrier where the therapeutic effect is desired.

244. PHARMACOLOGY: ANSWER: I
(DiPalma, 3/e, pp 414-415, 490-495) Menadione, the water-soluble form of vitamin K, should not be given to infants because of the high incidence of hemolysis and jaundice.

245. PHARMACOLOGY: ANSWER: H
(DiPalma, 3/e, pp 414-415, 490-495) Alpha-tocopherol, or vitamin E, is relatively nontoxic and has antioxidant properties, e.g., preserving intracellular components such as ubiquinone.

246. BIOCHEMISTRY: ANSWER: C
(Gilman, 7/e, p 1096) Methotrexate is often referred to as a folic acid or folate antagonist.

247. BIOCHEMISTRY: ANSWER: A
(Stryer, 3/e, p 626) Sulfisoxazole is a sulfa drug, an analogue of *p*-aminobenzoic acid.

248. BIOCHEMISTRY: ANSWER: E
(Stryer, 3/e, p 251) Warfarin is an anticoagulant related to vitamin K_1 which is required for normal clotting.

249. GENETICS: ANSWER: E
(Gelehrter, chap 8. Thompson, 5/e, chap 9) Offspring of translocation 21/21 carriers should in theory all have Down syndrome, although in practice some carriers have had normal children.

250. GENETICS: ANSWER: C
(Gelehrter, chap 8. Thompson, 5/e, chap 9) The empiric risk for parents with a trisomy 21 child is 1/100 for a second child with chromosomal aneuploidy.

251. GENETICS: ANSWER: D
(Gelehrter, pp 171-189. Thompson, 5/e, pp 201-214) The 21/21 translocation in this question is not associated with an additional normal 21 chromosome, as indicated by the chromosome number of 45.

252. GENETICS: ANSWER: A
(Gelehrter, pp 171-189. Thompson, 5/e, pp 201-214) In this question, there is a normal 21 chromosome in addition to the 21/21 translocation chromosome. As a result, this person has Down syndrome rather than being a balanced translocation carrier with a risk for chromosomally abnormal offspring.

253. PATHOLOGY: ANSWER: C
254. PATHOLOGY: ANSWER: C
255. PATHOLOGY: ANSWER: D
(Robbins, 4/e, pp 724-730) Auer rods are often prominent in the hypergranular promyelocytes of acute promyelocytic leukemia since they are formed from the abnormal azurophilic granules. Myeloblasts predominate in acute myeloblastic leukemia (AML) and, therefore, only a few granules, or occasional Auer rods, are present. Acute promyelocytic leukemia is associated with a short course and widespread petechiae and ecchymoses, cutaneous or mucosal, from disseminated intravascular coagulation. The M1 and M3 classes refer to the French-American-British (FAB) classification of AML: M1 is AML in which myeloblasts predominate and M3 is acute promyelocytic leukemia with promyelocytes numerous. Myeloperoxidase is present in both, especially in M3. AML occasionally follows chemotherapy and radiotherapy for Hodgkin's disease.

Chronic lymphocytic leukemia occurs most frequently after the age of 50 (90 percent of cases). It is associated with long survival in many cases and the few symptoms are related to anemia and the absolute lymphocytosis of small mature cells. Splenomegaly may be noted. Some patients are asymptomatic.

256. PHARMACOLOGY: ANSWER: A
(DiPalma, 3/e, pp 141-142) Acetylcholine is synthesized from acetyl-CoA and choline. Choline is taken up into the neurons by an active transport system. Hemicholinium blocks this uptake, depleting cellular choline, so that synthesis of acetylcholine no longer occurs.

257. PHARMACOLOGY: ANSWER: D
(DiPalma, 3/e, p 151) Hexamethonium is also a competitive inhibitor of acetylcholine, but this drug is specific for the nicotinic receptor in the ganglia. Dexamethonium, with 10 carbons instead of 6, is specific for nicotinic receptors at the neuromuscular junction. Hexamethonium causes hypotension by preventing sympathetic constriction of the blood vessels.

258. PHARMACOLOGY: ANSWER: D
(DiPalma, 3/e, pp 153-155) Muscarine, an alkaloid from certain species of mushrooms, is a muscarinic receptor agonist. The compound has toxicologic importance; muscarine poisoning will produce all the effects associated with an overdose of acetylcholine, e.g., bronchoconstriction, bradycardia, hypotension, excessive salivary and respiratory secretion, and sweating. Poisoning by muscarine is treated with atropine.

259. PATHOLOGY: ANSWER: B
260. PATHOLOGY: ANSWER: E
261. PATHOLOGY: ANSWER: A
(Anderson, 9/e, pp 1757, 1759-1760, 1762-1763, 1772-1774, 1794-1795) Lichen planus is an acute or chronic inflammation of the skin and mucous membranes characterized by flat-topped, pruritic, purple papules on the skin and white oral papules. Histologically, lesions show acanthosis with elongated, saw-toothed rete ridges and liquefaction degeneration of the basal layer, often with necrotic basal cells in the papillary dermis, which form colloid or Civatte bodies. There is usually a bandlike lymphocytic infiltrate along the dermoepidermal junction.

Verruca vulgaris, the common wart, is one of the human papillomavirus (HPV 1, 2, 3, 4) infections of the papovavirus group. It may occur anywhere but most commonly appears on the hands and fingers of schoolchildren as firm 1- to 10-mm papules. Histologically, the lesion shows a papillary acanthosis and cells of the stratum granulosum often display perinuclear vacuoles (koilocytosis). Intranuclear viral particles are seen by EM. A characteristic finding is of "reddish-brown" dots (thrombosed capillary loops) seen with a hand lens on the surface.

Acanthosis nigricans occurs most typically in flexures (axilla, groin, anogenital region) and is usually associated with benign or malignant conditions. The juvenile type is autosomal dominant with variable penetrance and occurs usually in association with obesity or endocrine disorders. The adult type (about 20 percent of all cases) often occurs with occult adenocarcinoma (usually gastric) but has occurred with oral contraceptive use. Folding of a hyperkeratotic epidermis with basal layer darkened with melanin is characteristic of acanthosis nigricans.

262. PHYSIOLOGY: ANSWER: C
263. PHYSIOLOGY: ANSWER: B
264. PHYSIOLOGY: ANSWER: A
(*Ganong, 15/e, pp 528-531; Guyton, 8/e, pp 106-111*) Cardiac output is a function of both stroke volume and heart rate. Alterations in contractility, which influences stroke volume, are termed *inotropic effects*, whereas alterations of rate are termed *chronotropic effects*. The Frank-Starling law of the heart states that the contractility of cardiac muscle is directly related to the initial length of the fiber. Thus, if cardiac muscle fibers are stretched by an increase in end-diastolic volume (acute volume overload), myocardial contractility increases (point C in the graph accompanying the question). In the presence of pericardial effusion, ventricular end-diastolic volume is depressed because of compression and so contractility decreases (point B).

In addition to this autoregulatory mechanism, myocardial contractility also is influenced by metabolic and neural factors. Sympathetic stimulation increases myocardial contractility (positive inotropic effect) and parasympathetic stimulation decreases it (negative inotropic effect). Of these two opposing factors, the sympathetic effect is the more potent and significant. Sympathetic stimulation is mediated by β-adrenergic receptors. Activation of these receptors results in increased cytoplasmic cyclic AMP, which mediates the inotropic effects. Drugs that inhibit degradation of cyclic AMP, such as caffeine, exert positive inotropic effects. Muscular exercise, by increasing sympathetic discharge, increases contractility as well as venous return (point A).

Hypoxia, hypercapnia, and acidosis all depress myocardial contractility so that for the same ventricular end-diastolic volume, contractility is reduced (point D).

265. GENETICS: ANSWER: A
266. GENETICS: ANSWER: A
267. GENETICS: ANSWER: B
268. GENETICS: ANSWER: C
269. GENETICS: ANSWER: D
(*Gelehrter, pp 27-47. Thompson, 5/e, pp 53-88*) Pedigree I has the vertical pattern suggestive of autosomal dominant inheritance. Although all affected individuals are male, X-linked inheritance is ruled out by the instance of male-to-male transmission. Affected individuals will have a 50 percent chance of affected offspring, regardless of sex. Pedigree II has the horizontal pattern of autosomal recessive inheritance and this is supported further by the presence of consanguinity. The proband has a 2 in 3 chance of being a carrier, and the consanguinity suggests that his wife is homozygous for the same RP alleles. The risk is thus 2/3 × 1/2 for the proband to contribute the recessive allele, while his wife has a 100 percent chance to do so, which results in a final probability of 1 in 3 that the child will be affected. This decreases to 1 in 9 if the proband marries his wife's unaffected sister, who also has a 2 in 3 chance of being a carrier.

270. MICROBIOLOGY: ANSWER: E
271. MICROBIOLOGY: ANSWER: C
272. MICROBIOLOGY: ANSWER: B
(Davis, 4/e, pp 201-228) Penicillin causes lysis of growing bacterial cells. Its antimicrobial effect stems from impairment of cell-wall synthesis. Because penicillin is bactericidal, the number of viable cells should fall immediately after introduction of the drug into the medium.

Sulfonamides are bacteriostatic—that is, they retard cell growth without causing cell death. Sulfonamides compete with para-aminobenzoic acid in the synthesis of folate; intracellular stores of folate are depleted gradually as the cells continue to grow.

The number of viable cells in a culture eventually will level off even if no antibiotic is added to the environment. A key factor in this phenomenon is the limited availability of substrate.

273. MICROBIOLOGY: ANSWER: D
274. MICROBIOLOGY: ANSWER: B
(Jawetz, 19/e, pp 458-462) The hepatitis B core antigen (HBcAg) is found within the nuclei of infected hepatocytes and not generally in the peripheral circulation except as an integral component of the Dane particle. The antibody to this antigen, anti-HBc, is present at the beginning of clinical illness. As long as there is ongoing HBV replication, there will be high titers of anti-HBc. During the early convalescent phase of an HBV infection, anti-HBc may be the only detectable serologic marker ("window phase") if HBsAg is negative and anti-HBsAg has not appeared.

275. MICROBIOLOGY: ANSWER: C
276. MICROBIOLOGY: ANSWER: B
(Jawetz, 19/e, pp 121-122) The complement-fixation (CF) test is a two-stage test. The first stage involves the union of antigen with its specific antibody, followed by the fixation of complement to the antigen-antibody structure. In order to determine whether complement has been "fixed," an indicator system must be employed to determine the presence of free complement. Free complement binds to the complexes formed when red blood cells (RBCs) are mixed with anti-RBC antibody; this binding causes lysis of the cells. Complement that has been "fixed" before addition of red blood cells and anti-RBC antibody cannot cause lysis.

277. BEHAVIORAL SCIENCE: ANSWER: D
(Williams, 3/e, pp 345-378) Medicare has two basic components. Part A is a compulsory insurance plan for persons 65 years or older who are entitled to benefits under the Social Security or Railroad Retirement Acts. Social Security payroll taxes provide the funds for inpatient diagnostic studies, hospital room and board costs, and home care and extended care services. Part B is a voluntary supplementary insurance program that pays for outpatient visits, diagnostic studies, doctors' fees, home health services, and certain medical equipment. Costs are shared between the individual and the Federal General Revenue Fund, except for a deductible charge and 20 percent of physicians' fees.

278. BEHAVIORAL SCIENCE: ANSWER: A
(Williams, 3/e, pp 345-378) Blue Cross is a prepaid, limited nonprofit commercial medical insurance plan to cover hospital costs. It has grown rapidly with the burgeoning costs of hospital care. It also has become increasingly popular as a negotiated health care benefit between industry and labor.

279. BEHAVIORAL SCIENCE: ANSWER: C

(Williams, 3/e, pp 345-378) The federal and state governments share the costs of Medicaid, although the program is administered by the states. The program provides medical care for low-income people of all ages and complements some of the provisions of Medicare. Medicare and Medicaid are providing limited and decreasing care. In spite of the four major payment plans discussed above, the major problems of equity, effectiveness, and economy still exist and are worsening. A very large segment of the population has no medical insurance coverage.

280. BEHAVIORAL SCIENCE: ANSWER: B

(Williams, 3/e, pp 345-378) Blue Shield is the same kind of payment mechanism as Blue Cross, extended to cover physicians' fees in hospitals as well as surgery and emergency care. Blue Cross, Blue Shield, Medicare, and Medicaid reimburse physicians and hospitals on a "reasonable cost" basis, which has continued to increase dramatically.

281. ANATOMY: ANSWER: D
282. ANATOMY: ANSWER: A
283. ANATOMY: ANSWER: C

(Hollinshead, 4/e, pp 681-685) The diaphragm possesses three principal hiatuses, shown in the diagram accompanying the question: the hiatus for the inferior vena cava (A), the esophageal hiatus (C), and the aortic hiatus (E). Potential diaphragmatic developmental defects include the foramen of Morgagni (B), just lateral to the xiphoid attachment of the diaphragm, and the pleuroperitoneal canal of Bochdalek (D), which is the most common site for congenital hernias. In the latter case, the pleuroperitoneal membranes fail to fuse to close the pleuroperitoneal canal—usually on the left side.

The inferior vena cava and, frequently, small branches of the right phrenic nerve pass through a hiatus (A) slightly to the right of the midline at the T8 level. The left phrenic nerve usually passes through the central tendon of the diaphragm on the left side to innervate the left hemidiaphragm from below.

The esophageal hiatus (C) just to the left of the midline at the T10 level transmits the esophagus, left and right vagus nerves, as well as the esophageal branches of the left gastric artery and vein. An acquired hiatus hernia usually is the consequence of a short esophagus or of a weakened esophageal hiatus. In such instances, a portion of the cardia and sometimes the fundus of the stomach slides upward through the diaphragm into the thorax. Radiographic examination will disclose a stomach that is constricted by the diaphragm, giving rise to the term ''hour-glass stomach.'' Hiatus hernia may be complicated by gastric regurgitation, esophagitis, and dysphagia.

284. PHARMACOLOGY: ANSWER: D

(DiPalma, 3/e, pp 411-412) Adverse reactions reported with the administration of thiazide diuretics (hydrochlorothiazide) include hypokalemia and hyperuricemia. In the nephron unit of the kidney, thiazides cause an increased excretion of sodium and chloride ions, and the loss of potassium ions. Chlorothiazide may also produce a loss of bicarbonate ion since it possesses some carbonic anhydrase inhibitory activity. The hypokalemia may be treated with the administration of potassium chloride or by the administration of a potassium-sparing diuretic such as spironolactone, triamterene, or amiloride. Hyperuricemia may be produced by the competition of the thiazide with the secretory pathway for uric acid. Although thiazides occasionally may induce hyperglycemia, they can be used in patients with diabetes mellitus if the patient is carefully evaluated and followed by the physician.

285. MICROBIOLOGY: ANSWER: E

(Jawetz, 19/e, pp 23-26) Flagella are organelles of motility. They are long, filamentous structures originating from a spherical basal body. A flagellum is composed of three parts: the filament, the hook, and the basal body.

286. PHYSIOLOGY: ANSWER: B

(Ganong, 15/e, pp 506-508) The PR interval in an ECG is measured from the beginning of the P wave, which reflects atrial depolarization, to the beginning of the QRS complex, which reflects initiation of ventricular depolarization. This interval normally ranges from 0.12 to 0.2 s and reflects the time required for atrial depolarization and conduction through the atrioventricular node. The PR interval is prolonged by vagal stimulation and hypokalemia and shortened by sympathetic stimulation.

287. BIOCHEMISTRY: ANSWER: D

(Stryer, 3/e, pp 164-169) Hemoglobin S, present in sickle-cell anemia, features a substitution of the hydrophobic Val residue for the hydrophilic Glu residue in hemoglobin A. In the three-dimensional configuration, this gives rise to a hydrophobic patch on the surface of the deoxygenated form of the molecule which favors aggregation in the presence of low concentrations of oxygen. Sickle cell anemia is an autosomal recessive hemoglobinopathy that occurs in approximately 1 in 500 births in the black population. It is caused by a single nucleotide substitution in codon 6 of the hemoglobin gene.

288. PATHOLOGY: ANSWER: B

(Anderson, 6/e, p 1433.) The histologic appearance of colloid storage goiter generally includes abnormally large colloid-filled follicles compressing the intervening small or normal-sized follicles that contain very little colloid. The epithelium of the follicles is predominantly flat cuboidal with occasional epithelial papillary structures protruding into the follicles. In primary hyperplasia with Graves' disease, the follicular epithelium is tall with papillary infoldings and peripheral vacuolation of the colloid. Riedel's struma appears as a marked fibrous tissue replacement of the normal thyroid histology. In Hashimoto's thyroiditis only remnants of thyroid follicles and epithelial cells are found in sheets of lymphocytes with germinal centers.

289. PHYSIOLOGY: ANSWER: D

(Berne, 2/e, pp 422-426) The magnitude of the QRS complex depends on the mass of tissue depolarized, the closeness of the ventricle to the recording electrode, and the relationship between the direction of ventricular depolarization and the orientation of the recording leads. Ventricular hypertrophy increases the magnitude of the QRS complex in all leads. When the mean electrical axis is horizontal, the size of the QRS complex in lead I increases while that in leads II and III diminishes. Although slowing of the heart rate may increase the force of contraction, it has no effect on the size of the QRS complex. In second-degree heart block, the ventricular beats that do occur have QRS complexes of normal shape.

290. PATHOLOGY: ANSWER: A

(Anderson, 9/e, pp 1898-1900) The triad of cystic bone lesions, precocious puberty, and patchy brownish skin pigmentation is known as Albright's syndrome. The bone lesions are those of fibrous dysplasia and apparently result from abnormal activity by the bone-forming mesenchyma. Packing of the medullary cavity by fibrous tissue that contains trabeculae of poorly mineralized fibrous bone is seen in the lesions. Recent reports have described cases of fibrous dysplasia in both males and females who also have had a wide variety of endocrine abnormalities, including hyperthyroidism, acromegaly, and Cushing's syndrome.

291. MICROBIOLOGY: ANSWER: B

(Baron, 3/e, pp 80-81) ATP is believed to be generated at three reaction points in the electron transport chain: the reductions of flavoprotein, cytochrome *b,* and cytochrome *c.* This phenomenon, demonstrated in experiments with mammalian mitochondria, can be expressed in terms of the relationship between the moles of ATP generated for each atom of oxygen consumed—the P/O ratio. In mammalian cells, the P/O ratio is 3; that is, there are three segments in the electron transfer chain in which there is a relatively large free energy drop. In bacteria, however, there appears to be only one or two of these segments. Loss of these phosphorylation sites as well as reactions that bypass these sites of ATP synthesis accounts for the lower P/O ratio in bacteria. Nevertheless, some bacteria, such as *Mycobacterium phlei,* do have P/O ratios of 3.

292. PATHOLOGY: ANSWER: A

(Robbins, 4/e, pp 173-183) The type of reaction in the question is a type 2 hypersensitivity reaction that is mediated by antibodies reacting against antigens present on the surface of cells, in this case blood group antigens or irregular antigens present on the donor's red blood cells. Type 2 hypersensitivity reactions result from attachment of antibodies to changed cell surface antigens or to normal cell surface antigens. Complement-mediated cytotoxicity occurs when IgM or IgG binds to a cell surface antigen with complement activation and consequent cell membrane damage or lysis. Blood transfusion reactions and autoimmune hemolytic anemia are examples of this form. Systemic anaphylaxis is a type 1 hypersensitivity reaction in which mast cells or basophils that are bound to IgE antibodies are reexposed to an allergen, which leads to a release of vasoactive amines that causes edema and broncho- and vasoconstriction. Sudden death can occur. Systemic immune complex reactions are found in type 3 reactions and are due to circulating antibodies that form complexes upon reexposure to an antigen, such as foreign serum, which then activates complement followed by chemotaxis and aggregation of neutrophils leading to release of lysosomal enzymes and eventual necrosis of tissue and cells. Serum sickness and Arthus' reactions are examples of this. Delayed-type hypersensitivity is type 4 and is due to previously sensitized T lymphocytes, which release lymphokines upon reexposure to the antigen. This takes time—perhaps up to several days following exposure. The tuberculin reaction is the best known example of this. T cell-mediated cytotoxicity leads to lysis of cells by cytotoxic T cells in response to tumor cells, allogenic tissue, and virus-infected cells. These cells have CD8 antigens on their surfaces.

293. BIOCHEMISTRY: ANSWER: C

(Stryer, 3/e, pp 412-420) As electrons move down the respiratory chain, protons are pumped out of the mitochondria generating a gradient of H^+ ions and of positive charge on the outside of the mitochondrial membrane. There are finite limits to the allowable size of these gradients, as well as several ways of using them up. For example, the F_o-F_1 ATPase carries protons back inside the mitochondria, simultaneously making ATP from ADP and P_i. In this example, the supply of ADP and P_i controls the rate of electron transport and thus the rate of oxygen consumption. The presence of an uncoupler allows protons to cross back generating only heat and preventing the synthesis of ATP.

294. MICROBIOLOGY: ANSWER: C

(Davis, 4/e, pp 687-697) All the listed diseases except Q fever are tick-borne. The rickettsia *Coxiella burnetii* causes Q fever and humans are usually infected by aerosol of a sporelike form shed in milk, urine, feces, or placenta of infected sheep, cattle, or goats. Lyme disease is caused by a spirochete, *Borrelia burgdorferi,* and produces the characteristic lesion erythema chronicum migrans (ECM). The etiologic agent of Rocky Mountain spotted fever is *R. rickettsia.* It usually produces a rash that begins in the extremities and then involves the trunk. Erhlichiosis is caused by *Erhlichia canis,* a rickettsia that is newly recognized as a human pathogen. It was previously considered only a pathogen in dogs. Ehrlichiosis produces the clinical picture in the question. Infection is transmitted by the brown dog tick, produces fever and leukopenia, and usually does not cause a rash. The organism infects monocytes and produces inclusion bodies in the phagosome, where it grows. *Francisella tularensis* is a small, gram-negative, nonmotile coccobacillus. Humans most commonly acquire the organism after contact with tissues or body fluid of an infected mammal or the bite of an infected tick.

295. BIOCHEMISTRY: ANSWER: C

(Stryer, 3/e, p 189) The Lineweaver-Burk plot is referred to as a double reciprocal plot. The y-axis data are expressed as $1/V$, and the x-axis data as $1/[S]$.

296. BIOCHEMISTRY: ANSWER: E

(Stryer, 3/e, pp 427-433, 437) The nonoxidative portion of the pentose pathway is completely reversible, allowing the synthesis of pentoses from glycolytic intermediates if the oxidative portion (which is where CO_2 is produced) is compromised. Under these conditions, however, an important source of NADPH is lost in direct proportion to the degree of the compromise. Glutathione is an oxidation-reduction buffer that has no direct role in carbon metabolism.

297. HISTOLOGY: ANSWER: E
298. HISTOLOGY: ANSWER: E

(Junqueira, 7/e, pp 330, 332. Ross, 2/e, p 354. Stevens, p 182) Bilirubin, a product of iron-free heme, is liberated during the destruction of old erythrocytes by the mononuclear macrophages of the spleen and, to a lesser extent, of the liver and bone marrow. The hepatic portal system brings splenic bilirubin to the liver, where it is made soluble for excretion by conjugation with glucuronic acid. Commonly, initial low levels of glucuronyl transferase in the underdeveloped smooth endoplasmic reticulum of hepatocytes in the newborn result in jaundice (neonatal hyperbilirubinemia); less commonly, this enzyme is genetically lacking. The ability of mature hepatocytes to take up and conjugate bilirubin may be exceeded by abnormal increases in erythrocyte destruction (hemolytic jaundice) or by hepatocellular damage (functional jaundice), such as in hepatitis. Finally, obstruction of the duct system between the liver and duodenum (usually of the common bile duct in the adult and rarely from aplasia of the duct system in infants) results in a backup of bilirubin (obstructive jaundice). However, obstruction of the cystic duct, while painful, will not interfere with the flow of bile from the liver to the duodenum.

299. PATHOLOGY: ANSWER: A

(Anderson, 9/e, pp 1853-1860) Malignant fibrous histiocytoma (MFH) is the current term used to designate sarcomatous growths of the deep soft tissues. This is principally a disease of the lower extremity and thigh in middle- to advanced-aged patients of both sexes, but it has also been reported as occurring in far-removed areas, such as the adventitia of the thoracic aorta and the ocular orbit. It has also occurred with some frequency years after irradiation. The tumor is composed of a background of spindle cells with varying

amounts of collagen (hence, fibrous) and scattered, giant, bizarre, xanthomatous and myoblastic-like cells (hence, histiocytic). In the past, many of these tumors were undoubtedly labeled as "pleomorphic liposarcomas" and "pleomorphic adult rhabdomyosarcomas," both of which must be differentiated from MFH. This does not imply, of course, that pleomorphic liposarcomas and adult rhabdomyosarcomas do not exist simply because a new term has been introduced, but rather that strict histologic criteria should be adhered to in order to exclude MFH—namely, that malignant myoblasts with cytoplasmic striations (rhabdomyosarcoma) and lipoblasts (liposarcoma) must be unequivocally demonstrated. The prognosis is poor for MFH.

300. PHARMACOLOGY: ANSWER: E

(DiPalma, 3/e, pp 123, 418) Prazosin and its close relative terazosin are unique because they block mainly α_1 receptors in contrast to phentolamine, which blocks both α_1 and α_2 receptors. The α_1-receptor selectivity permits the normal norepinephrine negative feedback on α_2 presynaptic receptors. Pindolol is a β blocker; minoxidil is a direct-acting vasodilator.

301. MICROBIOLOGY: ANSWER: C

(Jawetz, 19/e, pp 314-316) Hairs infected with *Microsporum canis* and *M. audouini* both fluoresce with a yellow-green color under Wood's light, while *Trichophyton rubrum, T. tonsurans,* and *Epidermophyton floccosum* do not. But *M. audouini* is an anthropophilic agent of tinea capitis, whereas *M. canis* is zoophilic. *M. canis* is primarily seen in children and is associated with infected cats or dogs.

302. PHYSIOLOGY: ANSWER: C

(West, 12/e, pp 524-525) Total ventilation is simply tidal volume times frequency. Some of this, however, is wasted ventilation of dead space. In the example provided in the question, the patient's and the machine's dead spaces must be added to determine the total dead space in each breath (of 1 L); the remainder is the alveolar ventilation. Thus, alveolar ventilation = (1000 mL – 200 mL – 50 mL) × 10/min = 7.5 L/min.

303. BEHAVIORAL SCIENCE: ANSWER: B

(Suinn, p 275) The characteristic symptoms of Korsakoff's psychosis are disorientation in time and place, confabulation, and anterograde amnesia. Anterograde amnesia is a specific cognitive deficit in which events that have just occurred are not recalled. Affected patients typically fabricate responses to questions about the forgotten occurrences. Inadequate nutritional habits (specifically, vitamin B deficiency) are the probable cause of Korsakoff's psychosis; hence the disorder frequently is associated with a history of chronic alcoholism.

304. BIOCHEMISTRY: ANSWER: B

(Stryer, 3/e, p 639) Early in starvation, the exclusive use of glucose by the brain requires that gluconeogenesis from amino acids be active. Urea is produced from the amino groups of the amino acids used. As starvation is prolonged, the brain adapts to the use of ketone bodies as fuel and less gluconeogenesis is required.

305. BIOCHEMISTRY: ANSWER: C

(Stryer, 3/e, pp 800-805) A typical inducible gene is turned off in the absence of the compound which is metabolized by the enzyme product of that gene. This mechanism involves the binding of a repressor protein to the operator region of the DNA of the gene, which precludes the binding of RNA polymerase. An inducer binds the repressor protein and causes it to dissociate from the operator region of the gene; RNA polymerase can now bind and the DNA is transcribed. The end result is an increase in the level of the inducible enzyme.

306. MICROBIOLOGY: ANSWER: E

(Balows, 5/e, pp 617-629) The determination of whether a patient is colonized or infected with *C. albicans* is not straightforward. Usually, isolation of the yeast from three sites is indicative of infection. Determination of antibody titer has little clinical utility. Detection of mannan or arabinitol has correlated with infection in some studies but not in others. Recent evidence suggests that detection of the protein antigen by latex agglutination may not signify disseminated candidiasis.

307. PHYSIOLOGY: ANSWER: C

(Ganong, 15/e, pp 88-89 and 102-103. Guyton, 8/e, pp 80-81 and 84; West, 11/e, pp 80-84) At rest, the difference in electrical potential across the membrane of the muscle cell (resting potential) is 90 mV. The nerve releases acetylcholine, which changes the ionic permeability of the muscle plasma membrane. Owing to the differences in concentration of sodium ions (high outside) and potassium ions (high inside) across the membrane, the increased permeability to ions gives rise to a sudden influx of sodium and efflux of potassium ions through the plasma membrane. This results in depolarization and subsequent repolarization of the muscle membrane.

308. BIOCHEMISTRY: ANSWER: D

(Stryer, 3/e, p 509) A form of vitamin B_{12} is a required cofactor in the enzyme methylmalonyl CoA mutase, which catalyzes the interconversion of methylmalonyl CoA and succinyl CoA. If intrinsic factor is deficient, vitamin B_{12} supply from the diet is decreased or lost. Methylmalonyl CoA is then converted to the corresponding acid, causing acidosis and resulting in excretion of methylmalonic acid.

309. PATHOLOGY: ANSWER: B

(Anderson, 9/e, pp 209-213) Heavy metal poisoning may occur via the respiratory route owing to contaminated inhalant and vapors. Such poisoning is usually industrially related, as with mercury (calomel workers), arsenic (pesticides), and lead (batteries and paints). Cadmium has been implicated in producing not only an acute form of pneumonia, but, with chronic exposure to small concentrations of cadmium vapors, diffuse interstitial pulmonary fibrosis and an increased incidence of emphysema as well. The "honeycomb" radiologic pattern is indicative of an interstitial fibrotic process and may be the result of repeated pneumonitis and bronchitis. Cadmium can also be found in tobacco smoke. Cobalt poisoning leads to myocardiopathy, mercury poisoning leads to renal tubular damage, and lead poisoning leads to liver necrosis and cerebral edema. Arsenic poisoning, in addition to carrying an increased risk of lung and skin cancer, may cause death by inhibition of respiratory enzymes and cardiac subendocardial hemorrhages complicated by gastroenteritis with shock.

310. MICROBIOLOGY: ANSWER: B

(Jawetz, 19/e, pp 325-327) Patients with a compromised cellular immune system, such as in AIDS, are susceptible to a wide variety of diseases, including infection with *Cryptococcus*. A brain abscess caused by *Cryptococcus neoformans* is not unusual in AIDS patients. Initial laboratory suspicion is usually aroused by

the presence of encapsulated yeast in the CSF. There also could be other microorganisms as well as noninfectious artifacts that superficially resemble yeast. While *C. neoformans* can be readily cultured, a rapid diagnosis can be made by detecting cryptococcal capsular polysaccharide in CSF or blood. Care must be taken to strictly control the test because rheumatoid factor may cross-react. Once the yeast is isolated, then specific stains as well as panels of assimilatory carbohydrates are available to definitively identify this organism as *C. neoformans*. The patient may also be infected with *Pneumocystis carinii*, but not in the central nervous system. *P. carinii* has recently been reclassified as a fungus.

311. CELL BIOLOGY: ANSWER: B

(Junqueira, 7/e, p 29. Ross, 2/e, pp 22-23. Widnell, pp 323-327) Freeze fracture is a preparative procedure in which tissue is rapidly frozen and then fractured with a knife. The fracture plane occurs through the hydrophobic central plane of membranes, which is the plane of least resistance to the cleavage force. The two faces are essentially the two interior faces of the membrane. They are described as the extracellular face (E face) and the protoplasmic face (P face). The P face is backed by the cytoplasm and in general contains numerous intramembranous particles. The E face is backed by the extracellular space and in general contains a paucity of intramembranous particles (see upper part of figure) compared with the P face (labeled with asterisks). Glycolipids and glycoproteins compose the glycocalyx that covers the cell membrane on its exterior surface.

312. PHARMACOLOGY: ANSWER: D

(DiPalma, 3/e, pp 44-47) The figure shows an elimination pattern that has two components. The upper linear portion represents distribution of cimetidine from the plasma to the tissues and is the alpha phase of elimination. True elimination is represented by the lower linear portion of the line (the beta phase), and this is what is used to determine the elimination half-life of the drug. This pattern typifies a two-compartment model. At 2 h after administration, the plasma concentration of cimetidine (Tagamet) is 1.0 µg/mL; at 2.3 h the concentration is 0.5 µg/mL. Therefore, the concentration of cimetidine decreased to one-half its initial value in 2.3 h–its half-life value. The half-life is independent of the drug concentration and dose administered. In addition, elimination usually occurs according to first-order kinetics: a linear relationship is obtained when the drug concentration is plotted on a *logarithmic* scale versus time on an *arithmetic* scale (a semilogarithmic plot).

313. PHARMACOLOGY: ANSWER: C

(DiPalma, 3/e, p 46) The fractional change in drug concentration per unit time is expressed by the elimination rate constant, k_e. This constant is related to half-life $\left(t_{1/2}\right)$ by the equation

$$k_e t_{1/2} = 0.693$$

The units of the elimination rate constant are time^{-1}, while the $t_{1/2}$ is expressed in units of time. By substitution of the appropriate value for half-life derived from the graph (beta phase) into the above equation, rearranged to solve for k_e, the answer is calculated as follows:

$$k_e = \frac{0.693}{t_{1/2}} = \frac{0.693}{2.3\ \text{h}} = 0.3\ \text{h}^{-1}$$

314. PHARMACOLOGY: ANSWER: D

(DiPalma, 3/e, pp 47-48) In the example given, a hypothetical plasma concentration of the drug at zero time (1.8 µg/mL) can be estimated by extrapolating the linear portion of the elimination curve (beta phase) back to zero time. The apparent volume of distribution is the volume of fluid into which the drug appears to distribute or the volume necessary to dissolve the drug and yield the same concentration as that found in plasma. The apparent volume of distribution (V_d) is calculated by

$$V_d = \frac{\text{Total amount of drug in the body}}{\text{Drug concentration plasma at zero time}}$$

$$V_d = \frac{300 \text{ mg}}{1.8 \text{ µg / ml}} = 111 \text{ L}$$

315. PHARMACOLOGY: ANSWER: A

(DiPalma, 3/e, pp 48-49) The total body clearance (CL_{total}) is the product of the volume distribution of the drug (V_d) and the elimination rate constant (k_e). It is an expression of the volume of the V_d cleared per unit time. The more rapidly a drug is cleared, the greater is the value of CL_{total}:

$$CL_{total} = V_d k_e = (111 \text{ L})(0.3 \text{ h}^{-1}) = 33.5 \text{ L / h}$$

316. MICROBIOLOGY: ANSWER: A

(Howard, pp 536-541) Evidence supporting the presence of fungal infection includes the clinical appearance of lesions and positive serologic reactions. However, detection of fungi in lesions, either by microscopic inspection or by culture, is the best evidence. Treating a specimen with 10% sodium hydroxide or potassium hydroxide hydrolyzes protein, fat, and many polysaccharides, leaving the alkali-resistant cell walls of most fungi intact and visible. With the availability of calcofluor white, however, KOH is not the optimum method for detecting and differentiating the structures of fungal cell walls. Calcofluor white is a nonimmunologic fluorescent stain that selectively stains fungi. Additionally, fluorescent antibody stains are available for the rapid diagnosis of fungal infection.

317. GENETICS: ANSWER: E

(Gelehrter, pp 49-65. Thompson, 5/e, pp 143-165) The Hardy-Weinberg term $p^2 + 2pq + q^2$ is useful for considering relative frequencies of genotypes even though these will be modified slightly by selection, migration, or inbreeding in actual populations. Since the incidence of achondroplasia is less than 1 in 10,000 births, the frequency of the abnormal allele (usually p is taken as the abnormal allele frequency in dominant diseases) is quite small (q is approximately equal to 1). For this reason, $p^2 < 2pq < q^2$ and homozygous abnormal patients will be extremely rare for most dominant diseases. Assortative mating (preferential mating among certain genotypes) is not uncommon in achondroplasia because of activities sponsored by organizations such as Little People of America. Such matings are the main source of homozygotes.

318. PHYSIOLOGY: ANSWER: D

(West, 12/e, pp 523-527) Changes in tidal volume (V_T) have a greater effect on alveolar ventilation ($\dot{V}_A$) than do equivalent changes in respiratory rate (RR) because of the contribution of the dead space. This relationship can be seen from the expression

$$\dot{V}_A = RR \, (V_T - \text{dead space})$$

Assuming, for example, that V_T = 500 mL, RR = 12/min, and dead space = 150 mL,

$$\dot{V}_A = 12 \, (500 - 150) = 4200 \text{ mL/min}$$

If V_T is doubled and RR halved,

$$V_T \times 2 = 100 \text{ mL}$$
$$RR \div 2 = 6 \, / \min$$

and

$$\dot{V} = 6 \, (1000 - 150) = 5100 \text{ mL} / \min$$

But if V_T is halved and RR doubled,

$$V_T \div 2 = 250 \text{ mL}$$
$$RR \times 2 = 24 \, / \min$$

and

$$\dot{V} = 24 \, (250 - 150) = 2400 \text{ mL} / \min$$

319. ANATOMY: ANSWER: B

(Hollinshead, 4/e, pp 232-234, 244) The dorsal antebrachial cutaneous nerve innervates the dorsum of the forearm and wrist, whereas the superficial radial nerve (of the hand) innervates the radial side of the dorsum of the hand (area B in the diagram accompanying the question). Because of considerable sensory overlap, a patient with radial nerve palsy usually experiences paresthesia in this region. However, a small area in the web space of the thumb, supplied exclusively by the radial nerve, provides a useful site for testing radial nerve sensory function. The dorsal branch of the ulnar nerve supplies the dorsum of the ulnar side of the hand (area D in the diagram) and the medial antebrachial cutaneous nerve, the lateral aspect of the forearm to the wrist (area C in the diagram). The lateral antebrachial cutaneous nerve (the extension of the musculocutaneous nerve into the forearm) supplies the lateral aspect of the forearm to the wrist (area A in the diagram). The dorsum of the tips of the index and middle fingers and a variable portion of the ring finger are supplied by the median nerve (area E in the diagram).

320. PATHOLOGY: ANSWER: B

(Robbins, 4/e, pp 574-576) Giant-cell arteritis (temporal arteritis), although not a major public-health problem, is an important disease to consider in the differential diagnosis of patients of middle to advanced age who present with a constellation of symptoms that may include migratory muscular and back pains (polymyalgia rheumatica), dizziness, visual disturbances, headaches, weight loss, anorexia, and tenderness over one or both of the temporal arteries. The cause of the arteritis (which may include giant cells, neutrophils, and chronic inflammatory cells) is unknown, but the dramatic response to corticosteroids suggests an immunogenic origin. The disease may involve any artery within the body, but involvement of the ophthalmic artery or arteries may lead to blindness unless steroid therapy is begun. Therefore, if clinically suspected, the workup to document temporal arteritis should be expedited and should include a biopsy of the temporal artery. Frequently, the erythrocyte sedimentation rate (ESR) is markedly elevated to values of 90 or greater. Whereas tenderness, nodularity, or skin reddening over the course of one of the scalp arteries, particularly the temporal, may show the ideal portion for a biopsy, it is important to recognize that temporal arterial segments may be segmentally uninvolved or not involved at all even when the disease is present.

321. BEHAVIORAL SCIENCE: ANSWER: A

(Simons, 3/e, pp 225-226) According to Freud, the superego is composed of two subsystems, the conscience and the ego ideal. The conscience refers to the set of internalized moral prohibitions that guides personal behavior. The ego ideal is the set of positive values and standards of correct behavior that a child internalizes and that the ego tries to emulate.

322. BIOCHEMISTRY: ANSWER: C

(Stryer, pp 568-569) A deficiency in steroid 11-hydroxylase would cause a deficiency of the adrenal steroids such as corticosterone, but would allow for an increase in the synthesis of testosterone. Clinically, there is an enlargement of the adrenals, presumably compensatory as a result of loss of negative feedback to the hypothalamo-hypophyseal system, and virilization of female patients from the increased circulating levels of testosterone.

323. BIOCHEMISTRY: ANSWER: A

(Stryer, 3/e, p 341) Lactose intolerance, characterized as bloating, cramping, and diarrhea after lactose ingestion, is due to excess fluid drawn into the gut by the undigested disaccharide in individuals who lack lactase, the enzyme that splits lactose into glucose and galactose. An inability to convert galactose to glucose results in galactosemia. Phosphorylation is not involved in lactose metabolism.

324. PHYSIOLOGY: ANSWER: C

(Berne, 2/e, pp 594-595. West, 12/e, pp 523-524) The helium dilution method can be used to determine FRC (which equals RV + ERV) since RV cannot be determined directly by spirometry. This method relies on the conservation of mass. A known volume and concentration (amount) of helium (which is inert in the body) is diluted by an unknown volume (the FRC of the person). The resulting helium concentration is measured, and since the amount of helium remains the same, the added volume — the FRC — can be calculated:

$$\text{Spirometer vol} \times \text{ initial \%He } = (\text{FRC} + \text{spirometer vol}) \times \text{final \%He}$$

$$12L \times 10\% = (\text{FRC} + 12) \times 8\%$$

$$\text{FRC} = 3L$$

325. MICROBIOLOGY: ANSWER: A

(Davis, 4/e, pp 869, 893-894) Recurrent severe infection is an indication for clinical evaluation of immune status. Live vaccines, including BCG attenuated from *Mycobacterium tuberculosis,* should *not* be used in the evaluation of a patient's immune competence because patients with severe immunodeficiencies may develop an overwhelming infection from the vaccine. For the same reason, oral (Sabin) polio vaccine is not advisable for use in such persons.

326. MICROBIOLOGY: ANSWER: C
327. MICROBIOLOGY: ANSWER: A
328. MICROBIOLOGY: ANSWER: C

(Jawetz, 19/e, pp 268-269) Mycoplasma pneumoniae causes a respiratory infection known as "primary atypical pneumonia" or "walking pneumonia." Although disease caused by *M. pneumoniae* can be contracted year round, thousands of cases occur during the winter months in all age groups. The disease, if untreated, will persist for 2 weeks or longer. Rare but serious side effects include cardiomyopathies and central nervous system complications. Infection with *M. pneumoniae* may be treated with either erythromycin or tetracycline. The organism lacks a cell wall and so is resistant to the penicillin and the cephalosporin groups of antibiotics.

Until recently, diagnostic tests have been of limited value. Up to 50 percent of cases may *not* show cold agglutinins, an insensitive and nonspecific acute-phase reactant. However, if cold agglutinins are present, a quick diagnosis can be made if signs and symptoms are characteristic. Complement fixation tests that measure an antibody to a glycolipid antigen of *M. pneumoniae* are useful but not routinely performed in most laboratories. Also, cross-reactions may occur. Culture of *M. pneumoniae,* while not technically difficult, may take up to 2 weeks before visible growth is observed. A DNA probe is available. It is an ^{125}I probe for the 16S ribosomal RNA of *M. pneumoniae*. Evaluations in a number of laboratories indicate that compared with culture it is highly sensitive and specific.

329. PATHOLOGY: ANSWER: A

(Robbins, 4/e, pp 598-601) The morphologic changes of clinical congestive heart failure cannot always be correlated with necropsy findings of the heart because there may be hypertrophy, dilatation, a combination of both, or even an absence of both. Many patients with long-standing congestive heart failure after decompensation will have hearts that are maximally dilated, with thinned and unusually soft myocardium rather than hypertrophic ventricular myocardium. This thinning of the myocardium occurs after a long period of compensatory hypertrophy and reflects a state in which the capacity of the myocardium to compensate has been exceeded. The first response of the myocardium to a demand for increased work (load) is to undergo hypertrophy according to Starling's law, leading to an increase in stroke volume. Eventually, this mechanism is exceeded under states of increased oxygen demand or demand for more cardiac output, and cardiac decompensation results, with the worst complication being acute pulmonary edema as a consequence of left ventricular failure.

330. PHARMACOLOGY: ANSWER: B

(DiPalma, 3/e, p 352) The usual electrocardiographic pattern of a patient receiving therapeutic doses of digitalis includes an increase in the PR interval, depression and sagging of the ST segment, and occasional biphasia or inversion of the T wave. Symmetrically peaked T waves are associated with hyperkalemia or ischemia in most cases. Shortening of the QT interval, rather than prolongation, is characteristic of digitalis treatment.

331. PHYSIOLOGY: ANSWER: C

(*Rose, 3/e, pp 211-215, 590-595. West, 12/e, pp 415-417*) Sweat normally contains about 40 to 60 meq of sodium per liter of fluid. Thus, approximately 100 meq of sodium will be lost from the extracellular fluid during the exercise period, and when the lost water is replenished, the extracellular fluid will become hypotonic. The hypotonic extracellular fluid will equilibrate with the intracellular fluid and make it hypotonic as well. Because the extracellular fluid volume is dependent on the amount of sodium, the loss of sodium will result in a decreased extracellular fluid volume and an increased intracellular fluid volume after water is replaced.

332. BEHAVIORAL SCIENCE: ANSWER: B

(*Counte, pp 65-68*) Low rates of compliance appear to result from defective relationships between patients and health care providers. The most crucial element in the physician-patient relationship appears to be the nature of role expectations that each has and the congruence and mutuality of such expectations. The exchange of information and facts, similarity of ages, social class differences, and patient rewards are relevant at times but appear to be of less importance than the congruence of expectations between the physician and the patient.

333. PATHOLOGY: ANSWER: D

(*Anderson, 9/e, pp 220-222*) Incomplete combustion of any carbon fuel will lead to accumulation of carbon monoxide gases. Oxygen deprivation results from the formation of carboxyhemoglobin, which displaces the normal oxyhemoglobin and interferes with oxygen exchange. In addition to small petechial hemorrhages of serosa and white matter of the cerebral hemispheres, patients who have lived for several days following exposure will have gross lesions in the brain that consist of bilateral hemorrhage and necrosis of the globus pallidus and hippocampus.

334. BIOCHEMISTRY: ANSWER: C

(*Stryer, p 503*) Leucine is exclusively ketogenic, metabolized only to acetyl CoA which cannot be converted to glucose in mammals. Glycine is glucogenic, while phenylalanine and isoleucine are both ketogenic and glucogenic.

335. MICROBIOLOGY: ANSWER: A

(*Balows, 5/e, pp 744-748*) Contaminated contact lens solutions have been recently implicated in serious infections of the eye caused by *Acanthamoeba*. The use of homemade saline is to be avoided; most commercial solutions, however, have preservatives in them. The organisms are difficult to isolate, but the differential criteria for the free-living ameba are both nutritional and morphologic.

336. PHYSIOLOGY: ANSWER: B

(*Berne, 2/e, pp 606-607. West, 12/e, pp 527, 559*) The volume of physiologic dead space (V_D) can be calculated using the Bohr equation:

$$V_D = \left(\frac{F_A CO_2 - F_E CO_2}{F_A CO_2} \right) V_T$$

V_T is tidal volume, $F_A CO_2$ is the fraction of CO_2 in alveolar gas, and $F_E CO_2$ is the fraction of CO_2 in mixed expired gas. $F_A CO_2$ can be approximated using the FCO_2 from a sample of end-expirator gas or by sampling

Pa_{CO_2}. In normal lungs, physiologic dead space equals anatomic dead space. In diseased lungs, physiologic dead space often exceeds anatomic dead space because of abnormalities of ventilation and blood flow.

337. PATHOLOGY: ANSWER: A
(Robbins, 4/e, pp 868-872) Hemorrhagic cobblestone appearance of the colon and small bowel may be seen in multiple states including inflammatory bowel disease, a term that can apply both to ulcerative colitis and regional enteritis (Crohn's disease). Other conditions that resemble cobblestoning of the mucosa of the bowel include multiple polyps such as occur in Gardner's syndrome, Turcot syndrome, familial polyposis, and multiple acquired polyps. Crohn's disease, however, differs from the others in that longitudinal ulcers may be present, yielding a long axis grooving, parallel to the long axis of the bowel. Such ulcers may also be seen in tuberculous enteritis; however, when inflammatory bowel disease is in the differential diagnosis, longitudinal ulcers are indicative of Crohn's disease.

338. BIOCHEMISTRY: ANSWER: C
(Stryer, 3/e, pp 354-355) Pyruvate kinase catalyzes the conversion of PEP to pyruvate with the generation of ATP. In muscle, this would cause fatigue. Since liver and brain pyruvate kinase are not known to be deficient, neither hypoglycemia nor retardation would be expected. You would expect high levels of fructose 1,6-bisphosphate (all steps toward it are reversible) and no effect on fructose metabolism.

339. PHARMACOLOGY: ANSWER: C
(DiPalma, 3/e, pp 48-49) The total body clearance (CL_{total}) can be calculated by the equation

$$CL_{total} = \frac{(0.693)(V_d)}{t_{1/2}}$$

where V_d is the apparent volume of distribution and $t_{1/2}$ is the half-life. V_d = 10 L (100 mg dose/10 mg/L plasma concentration). Therefore,

$$CL_{total} = \frac{(0.693)(10 \text{ L})}{7 \text{ h}}$$

$$CL_{total} = 0.99 \text{ L / h or } 16.5 \text{ mL / min}$$

CL_{total} represents the sum of clearance from all participating organs in the body, but mainly liver and kidney; that is,

$$CL_{total} = CL_{liver} + CL_{kidney}$$

Since clearance of the parent drug in the urine (unmetabolized) was 8.25 mL/min, this occurred via the kidney. Therefore,

$$CL_{total} - CL_{kidney} = CL_{liver}$$

$$16.50 \text{ mL / min} - 8.25 \text{ mL / min} = 8.25 \text{ mL / min}$$

or 50 percent of drug elimination is attributed to metabolism.

340. BIOCHEMISTRY: ANSWER: D

(Stryer, 3/e, pp 458-459) Glucagon is a polypeptide hormone whose action raises blood glucose levels, both by increasing the mobilization of glycogen and by increasing gluconeogenesis in the liver. Glycogen phosphorylase is the controlled enzyme in the mobilization of glycogen. HMG CoA reductase is involved in the synthesis of cholesterol; citrate lyase and acetyl CoA carboxylase are involved in the synthesis of fatty acids. Glucose-6-phosphatase is required for the production of glucose but is constitutive and not regulated by glucagon.

341. HISTOLOGY: ANSWER: D

(Erlandsen, pp 113, 115. Junqueira, 7/e, p 290. Ross, 2/e, pp 425, 429. Stevens, p 164. Wheater, 2/e, pp 209, 213) The photomicrograph shows the pylorus of the stomach. The pylorus differs from the fundus of the stomach in the length of the pits of the glands compared to the length of the gland. In the fundus there are short pits and long glands (pit:gland ratio of about 1:4) compared with a pit:gland ratio of 1:2 in the pylorus. There is also an absence of parietal cells in the pylorus.

342. BIOCHEMISTRY: ANSWER: B

(Stryer, 3/e, pp 706-707) The sigma factor is an element of RNA polymerase that allows it to recognize specific regions of DNA as promoter sites, allowing transcription in response to recognition of those sites. The separation of the DNA strands is accomplished by a helicase. The sigma factor dissociates soon after transcription begins, and is thus not required for elongation or release. The inducer is what binds to repressor proteins.

343. PATHOLOGY: ANSWER: C

(Robbins, 4/e, pp 764-765) Many pathologic pulmonary changes can be found in the lungs of patients who expire under conditions of progressive, unexplained dyspnea, fatigue, and cyanosis. These changes range from pulmonary fibrosis to hypersensitivity pneumonitis and to recurrent, multiple pulmonary emboli. Furthermore, traditional hospital treatment modalities for progressive pulmonary deterioration (including high oxygen delivery, overhydration, lack of pulmonary ventilation, irregular ventilation by mechanical respiratory assistance [PEEP], and superimposed nosocomially acquired pneumonitis) can complicate pulmonary pathologic findings. However, unremitting, progressive dyspnea, cyanosis, and fatigue in a young woman should suggest the diagnosis of primary pulmonary hypertension. Pulmonary vascular sclerosis is always associated with pulmonary hypertension primary or secondary to other states, such as emphysema and mitral stenosis.

344. MICROBIOLOGY: ANSWER: E

(Ash, 3/e, pp 65-67) Primary amebic meningoencephalitis, associated with swimming in freshwater lakes, is a rare, usually fatal disease. The causative agent, *Naegleria fowleri*, gains access to the central nervous system by being forced under pressure through the nasal mucosa covering the cribriform plate. Diagnosis is made by observation of the motile amebae in a wet mount of the cerebrospinal fluid. The organism can be cultured, and the indirect fluorescent antibody test can confirm the identification.

345. PATHOLOGY: ANSWER: C

(Robbins, 4/e, pp 110-111) Fat embolism syndrome can supervene as a complication within 3 days following severe trauma to the long bones. However, the pathogenesis must be regarded as unknown because simple entrance of microglobular fat into the circulation as a result of damage to small vessels in marrow tissue occurs in over 90 percent of patients with trauma and bone fracture, yet the syndrome occurs in only a minority of such patients. Laboratory and clinical findings can simulate those of intravascular coagulopathy, which may be a component of fat embolism syndrome, with a major difference of split

products of fibrin seen mainly in intravascular coagulopathy. Plasma levels of free fatty acids are elevated in fat embolism and may contribute to pulmonary vascular alterations. At autopsy fat material can be demonstrated in fat stains of frozen sections of lung, brain, and kidney in patients who had the syndrome.

346. MICROBIOLOGY: ANSWER: B

(Balows, 5/e, pp 738-739) As in the case of this archeologist, there have been a number of cases of both visceral and cutaneous leishmaniasis in Desert Storm veterans returning from Saudi Arabia. The CDC has recommended that all returning veterans not donate blood, although the risk of transfusion-related leishmaniasis appears low. The kinetoplast aids in differentiating amastigotes from other similar organisms that may be found in macrophages.

347. PHARMACOLOGY: ANSWER: C

(DiPalma, 3/e, pp 405-406) Lovastatin decreases cholesterol synthesis in the liver by inhibiting HMG-CoA reductase, the rate-limiting enzyme in the synthetic pathway. This results in an increase in LDL receptors in the liver, thus reducing blood levels for cholesterol. The intake of dietary cholesterol must not be increased, as this would allow the liver to use more exogenous cholesterol and defeat the action of lovastatin.

348. PHYSIOLOGY: ANSWER: B

(Berne, 2/e, pp 424-429. West, 12/e, pp 189-196) The normal PR interval is between 0.12 and 0.2 s. In this ECG, the PR interval is 0.25 s and thus the patient has a first-degree heart block. The heart rate is 40 beats per minute, which is a bradycardia; a tachycardia is a heart rate above 100 beats per minute. There is no sign of any rhythm disturbance such as sinus arrhythmia or atrial flutter.

349. MICROBIOLOGY: ANSWER: D

(Ash, 3/e, pp 74-77) The infection rate with *Giardia lamblia* in male homosexuals has been reported to be from 21 to 40 percent. These high prevalence rates are probably related to three factors: the endemic rate, the sexual behavior that facilitates transmission (the usual barriers to spread have been interrupted), and the frequency of exposure to an infected person.

350. PHYSIOLOGY: ANSWER: E

(West, 12/e, pp 524-525) Minute volume, or minute ventilation, is the tidal volume times the breathing frequency. This includes both alveolar ventilation ($\dot{V}_A$) and dead space ventilation ($\dot{V}_D$): minute ventilation = $\dot{V}_A + \dot{V}_D$. Dead space ventilation is the dead space volume (V_D) times the breathing frequency (f): $\dot{V}_D = V_D \times f$. Substitution into the equations yields

$$\text{Alveolar ventilation} = \text{alveolar volume} \times \text{breathing frequency}$$

$$= 250 \times 20 \text{ b / min}$$

$$= 5000 \text{ mL / min}$$

$$\text{Tidal volume} = \text{alveolar volume} + \text{dead space}$$

$$\text{Alveolar volume} = \text{Tidal volume} - \text{dead space}$$

$$= 400 - 150$$

$$= 250$$

351. BIOCHEMISTRY: ANSWER: B

(Stryer, pp 637-638) Muscles do not have glucose-6-phosphatase and thus do not contribute to blood glucose. A deficiency in liver glucose-6-phosphatase will compromise glucose production from gluconeogenesis and from glycogen mobilization. Lack of phosphorylase will prevent glycogen mobilization. Lack of amylo-1,6-glucosidase will prevent the debranching of glycogen molecules, resulting in the mildest disease.

352. PATHOLOGY: ANSWER: C

(Robbins, 4/e, pp 353-355, 867-869, 886-889) Early stages of ulcerative colitis (UC) may be indistinguishable from gastroenteritis caused by *Salmonella choleraesuis* and *S. typhimurium*. In early stages, both diseases may show histologically a dense mononuclear inflammatory infiltrate in the lamina propria, occasional crypt abscesses, and mucosal edema and congestion. Even the respective clinical symptoms and colon x-ray changes may be similar, although marked vomiting should point to food poisoning. Salmonellae have been the cause of outbreaks and epidemics of acute gastroenteritis, and the cause has often been found to be contaminated fowl that has been insufficiently cooked to inactivate endotoxins.

353. HISTOLOGY: ANSWER: D

(Junqueira, 7/e, pp 459-461. Wheater, 2/e, p 303. Yen, 3/e, pp 837-839) The patient described in this question is probably pregnant. The delay in menstruation coupled with the presence of basophilic cells in a vaginal smear is a clue. Ovulation is the midpoint of the cycle and should be more than a few days away. She is relatively young for the onset of menopause and there are no other symptoms. The vaginal epithelium varies little with the normal menstrual cycle. Exfoliative cytology can be used to diagnose cancer and to determine if the epithelium is under stimulation of estrogen and progesterone. The presence of basophilic cells in the smear with the Pap staining method would indicate the presence of both estrogen and progesterone. The data suggest the maintenance of the corpus luteum (i.e., pregnancy).

354. PHYSIOLOGY: ANSWER: D
355. PHYSIOLOGY: ANSWER: D
356. PHYSIOLOGY: ANSWER: E

(Ganong, 15/e, pp 669-671, 681-682. Guyton, 8/e, pp 335-339) Net acid excretion is equal to the sum of the amount of titratable acids plus the amount of the ammonium ions minus the amount of bicarbonate ions contained in the volume of urine produced per day.

$$\text{Net acid excretion} = \left([\text{titratable acids}] + \left[NH_4^+ \right] - \left[HCO_3^- \right] \right) \times \text{urine volume per day}$$

$$= \left(10 \text{ meq / L} + 20 \text{ meq / L} - 4 \text{ meq / L} \right) \times 1.5 \text{ L / day}$$

$$= 39 \text{ meq / day}$$

Acid is excreted in urine bound to titratable acids, principally phosphates, and bound to ammonia as NH_4^+. If bicarbonate is present in urine, its amount must be subtracted because a bicarbonate ion excreted in the urine means that a hydrogen ion was left behind in the body. Conversely, a net hydrogen ion excreted means that a bicarbonate ion was left behind in the body. Thus, net acid excretion is equal to new bicarbonate formation.

The normal fixed acid production is 70 to 100 meq/day. This is eliminated by the kidneys. This patient excreted only 39 meq H^+ per day and is therefore conserving hydrogen ions. He is also excreting HCO_3^- (retaining H^+) and must therefore be in a state of respiratory alkalosis. In respiratory and metabolic acidoses, the excretion of H^+ ion by the kidney is increased above normal and no bicarbonate ion is excreted in the urine.

357. PATHOLOGY: ANSWER: E

(*Robbins, 4/e, p 1078*) A middle-aged patient is highly unlikely to have either of the predominantly childhood tumors nephroblastoma (Wilms' tumor) or mesoblastic nephroma (benign hamartoma). Mesoblastic nephroma, which may be seen in the first year of life, has caused difficulty in differential diagnosis from Wilms' tumor in children. Acute pyelonephritis features signs of acute infection with flank pain, pyuria, fever, and a high bacterial colony count in urine. Renal cell carcinoma is unlikely to cause hematuria until far advanced with invasion of the collecting system. Urothelial renal pelvic tumors cause hematuria early, even when quite small. They form 5 to 10 percent of primary renal tumors and range from apparently benign papillomas to papillary or anaplastic carcinomas. There may be multicentric involvement of ureters or bladder. Diagnosis is by x-ray and cytologic examination of at least three voided urine specimens; malignant cells are not found if the ureter is obstructed by tumor, or if the cells are degenerate or mildly atypical, as in papilloma. Prognosis is not very good for high-grade infiltrating tumors and is very poor for the squamous cell variant (about 15 percent of pelvic tumors); therefore, early diagnosis is paramount.

358. MICROBIOLOGY: ANSWER: B

(*Jawetz, 19/e, pp 117, 143*) Interleukin 1 is a protein produced by macrophages that has three biologically active forms: IL-1α, β, and γ. Its functions include activation of B cells and stimulation of helper and cytotoxic T cells. Its activity is not histocompatibility-restricted.

359. BIOCHEMISTRY: ANSWER: C

(*Stryer, pp 975-978*) The slight rise in blood glucose following epinephrine injections demonstrates that liver glycogen storage and mobilization are normal. A fructose-1,6-bisphosphatase deficiency would produce these symptoms, but glucagon would then restore euglycemia.

360. MICROBIOLOGY: ANSWER: A

(*Ash, 3/e, pp 106-113*) The febrile paroxysms of *Plasmodium malariae* malaria occur at 72-h intervals; those of *P. falciparum* and *P. vivax* malaria occur every 48 h. The paroxysms usually last 8 to 12 h with *P. vivax* malaria but can last 16 to 36 h with *P. falciparum* disease. In *P. vivax*, *P. ovale*, and *P. malariae* infections, all stages of development of the organisms can be seen in the peripheral blood; in malignant tertian (*P. falciparum*) infections, only early ring stages and gametocytes are usually found.

361. PATHOLOGY: ANSWER: E

(Fitzpatrick, 2/e, pp 360-361. Robbins, 4/e, pp 1198-1199) The biopsy shows infiltration of the nipple by large cells with clear cytoplasm, which is diagnostic of Paget's disease. These cells are usually found both singly and in small clusters in the epidermis. Paget's disease is always associated with (in fact begins with) an underlying intraductal carcinoma that extends to infiltrate the skin of nipple and areola. Paget cells may resemble the cells of superficial spreading melanoma, but they are PAS-positive diastase-resistant (mucopolysaccharide- or mucin-positive), unlike melanoma cells. Eczematous dermatitis of the nipples is a major differential diagnosis but is usually bilateral and responds rapidly to topical steroids. Paget's disease should be suspected if "eczema" persists more than 3 weeks with topical therapy. Paget's disease occurs mainly in middle-aged women but is unusual. In Paget's disease of the vulvar-anal-perineal region, there is very rarely underlying carcinoma. Mammary fibromatosis is a rare, benign, spindle cell lesion affecting women in the third decade. Clinically, it may mimic cancer with retraction or dimpling of skin. It should be treated by local excision with wide margins since there is risk of local recurrence.

362. PATHOLOGY: ANSWER: A

(Robbins, 4/e, pp 1187-1191) The spectrum of benign breast disease includes fibrocystic disease, which is probably a misnomer; adenosis, both sclerosing and microglandular; intraductal papillomas and papillomatosis; apocrine metaplasia; fibrous stromal hyperplasia; and hyperplasia of the epithelial cells lining the ducts and ductules of the breasts. At one time or another each of the above was considered to be a forerunner of carcinoma; however, with extensive studies in the literature, none of these has been shown to necessarily correlate with a greater risk of developing carcinoma with the exception of epithelial hyperplasia. With any of the features, but especially epithelial hyperplasia, adding a positive family history of breast cancer in a sibling, mother, or maternal aunt markedly increases the risk for developing carcinoma of the breast in the given patient. Owing to the advances and technology of xeromammography, there has been an increased interest in calcifications, which are markers for carcinoma of the breast. These calcifications, however, do not necessarily occur within the cancerous ducts themselves and can be found frequently in either adenosis adjacent to the carcinoma or even in normal breast lobules in the region. Stipple calcification as seen by xeromammography is regarded as an indication for a biopsy of the region by some workers.

363. PATHOLOGY: ANSWER: D

(Robbins, 4/e, pp 204-207, 579) The constellation of Raynaud's phenomenon, acral sclerosis, and fibrotic tightening of the muscles of facial expression should raise the specter of progressive systemic sclerosis (scleroderma), a multisystem disease that involves the cardiovascular, gastrointestinal, cutaneous, musculoskeletal, pulmonary, and renal systems through progressive interstitial fibrosis. Small arterioles in the forenamed systems show obliteration caused by intimal hyperplasia accompanied by progressive interstitial fibrosis. Evidence implicates a lymphocyte overdrive of fibroblasts to produce an excess of rather normal collagen. Eventually, myocardial fibrosis, pulmonary fibrosis, and terminal renal failure ensue. Over half of all patients have dysphagia with solid food caused by the distal esophageal narrowing in the disease.

364. BIOCHEMISTRY: ANSWER: C

(Stryer, p 564) HMG CoA reductase is the committed and most regulated step in the synthetic pathway leading to cholesterol. It makes sense to attempt to inhibit this activity with a competitive or irreversible inhibitor, which would most likely be a structural analogue.

365. CELL BIOLOGY: ANSWER: C

(Junqueira, 7/e, pp 41-43. Stevens, p 19) In adrenoleukodystrophy there is an absence or deficiency in the peroxisomal enzymes that process long-chain fatty acids. There is a buildup of lipid storage in the adrenal cortex and brain, and this leads to adrenocortical failure and dementia, respectively. With the failure of the adrenal cortex and decreased production of adrenocortical hormones, one would expect increased ACTH (adrenocorticotropin) levels rather than decreased concentrations in the blood.

366. BIOCHEMISTRY: ANSWER: B

(Stryer, 3/e, p 55) Trypsin catalyzes the hydrolysis of the peptide bond *C*-terminal to a basic residue, which in this case is Lys. Gly must be the *C*-terminal to Lys and the only remaining place for Ala is the *N*-terminal.

367. ANATOMY: ANSWER: B

(Langman, 6/e, p 61. Moore, 4/e, pp 65, 87) The first month of embryonic development generally is concerned with cleavage, formation of the germ layers, and establishment of the embryonic body. Formation of most internal organs occurs during the second month, the period of organogenesis. The period from the ninth week to the end of intrauterine life, known as the fetal period, is characterized by maturation of tissues and rapid growth of the fetal body.

368. HISTOLOGY: ANSWER: B

(Junqueira, 7/e, pp 289, 451, 459. Ross, 2/e, pp 446-447, 450-451, 456-457, 696-697, 702-703. Stevens, pp 155, 174-175, 325, 327) The histologic section in the photomicrograph accompanying the question shows two distinctly different types of epithelium. The esophageal mucosa on the left is nonkeratinized, stratified squamous epithelium overlying a fibrovascular submucosa and smooth muscle. The gastric mucosa on the right is simple columnar epithelium with simple glands, overlying submucosa and smooth muscle. Skin is keratinized. Both the stomach and duodenum are simple columnar epithelium. The simple columnar epithelium of the distal-most rectum contains only mucous cells. The cervical mucosa contains extensive cervical glands and the vaginal epithelium is keratinized.

369. ANATOMY: ANSWER: B

(Noback, 3/e, p 318) The patient discussed in the question demonstrates Wallenberg's syndrome. Involvement of the motor and sensory divisions of the left trigeminal nerve with left cerebellar signs places the lesion in the left lateral portion of the midpons. The spinothalamic tract is involved, producing loss of pain and temperature sensation from the same side of the face and the opposite side of the body. Wallenberg's syndrome is frequently the result of occlusion of the circumferential branch of the basal artery.

370. PHARMACOLOGY: ANSWER: D

(DiPalma, 3/e, pp 432-433) Elderly people are apt to have low-reserve kidneys owing to renal artery stenosis. Inhibitors of angiotensin converting enzyme depress renal function and cause proteinuria even in patients with normal kidneys. In the elderly this toxicity is more manifest and may precipitate acute renal failure. The other toxicities listed in the question do not occur at a higher frequency in the elderly.

371. BIOCHEMISTRY: ANSWER: A

(Stryer, pp 357-358) Normally, fructose is converted to fructose-1-phosphate, and aldolase splits this molecule into DHAP and glyceraldehyde for further metabolism. If aldolase fails to recognize fructose-1-phosphate, it piles up, interfering with glucose metabolism and damaging the liver. Enzymes B and C are in the regular glycolytic pathway and enzyme D is important in gluconeogenesis.

372. BIOCHEMISTRY: ANSWER: B

(Stryer, pp 475-476)   In the absence of carnitine, palmitic acid, oleyl-CoA, and palmityl-CoA cannot traverse the mitochondrial membrane for oxidation. Only the fatty acyl carnitine will be oxidized.

373. PHARMACOLOGY: ANSWER: C

(DiPalma, 3/e, p 255)   The parent compound of the phenothiazines, a class of agents primarily used as antipsychotics, consists of two benzene rings linked by a sulfur and nitrogen atom to form a three-ring structure as shown in the question. It is similar to anthracene, except that C9 is replaced by sulfur, C10 is replaced by nitrogen, and various groups (R_1 and R_2) are substituents on the nitrogen and C3 of the generic phenothiazine molecule, as depicted in the following structures:

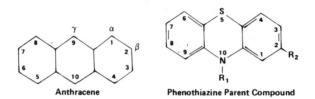

Anthracene   Phenothiazine Parent Compound

If R_2 is chlorine, the potency for depressing motor activity is increased. If R_1 is a piperazine or a piperazinyl group, the parent compound becomes the most potent of the phenothiazine antipsychotic drugs. Tricyclic antidepressants have an ethylene bridge as a replacement for the sulfur and have no substitutions at R_2.

374. BIOCHEMISTRY: ANSWER: B

(Stryer, pp 800-802)   In a typical inducible gene, a repressor protein binds near the start of the coding sequence, preventing access by RNA polymerase and preventing transcription of mRNA. The inducer binds with the repressor, causing it to release the DNA and allowing the synthesis of mRNA.

375. PATHOLOGY: ANSWER: E

(Robbins, 4/e, pp 1005-1008. Schwarz, p 917)   Any tumor of the pancreas seen to have an organoid (endocrine-like) pattern histologically, even in frozen section, should arouse suspicion of an islet cell tumor, a carcinoid, or a component tumor of multiple endocrine neoplasia. The pathologist may be alert to this possibility if appropriate clinical information relating the patient's symptoms accompanies the biopsy specimen. Electron microscopy will show specialized types of electron-dense core granules ("neurosecretory" granules) in the cytoplasm in the presence of tumors of the amine precursor uptake decarboxylation class (APUDomas). Islet cell tumors may contain alpha (glucagon), beta (insulin), delta (somatostatin), and pp (pancreatic polypeptide) dense-core granules. Direct staining of the hormones can be accomplished with immunoperoxidase, which contains the specific antibody to the hormone being sought and forms rust-brown granules that can be seen with the ordinary light microscope.

376. BIOCHEMISTRY: ANSWER: E

(Stryer, pp 163-168)   Sickle cell anemia is an autosomal recessive hemoglobinopathy that occurs in approximately 1 in 500 births in the black population. It is caused by a single nucleotide substitution in codon 6 of the hemoglobin gene. Patients suffering from sickle cell anemia make an abnormal globin β-chain, in which a glutamic acid residue is replaced by a valine residue. This results in a hydrophobic patch on the surface of the deoxygenated form of the molecule, and a tendency to polymerize and precipitate, causing the red blood cells to be distorted and to clog capillaries.

377. BEHAVIORAL SCIENCE: ANSWER: D

(*Lerner, 1986. pp 334-343*) Erik Erikson's stage theory of psychosocial development provides descriptions of personality development that are consistent with a life-span perspective. It has served as a major basis for research on human development. Erikson views the individual as having to develop the capacities (the ego functions) to meet the expectations of society. At each of the eight stages of psychosocial development there is an accompanying ''ego crisis.'' These stages are trust versus mistrust (birth to 1 1/2 years), autonomy versus shame or doubt (1 1/2 to 3 years), initiative versus guilt (3 to 6 years), industry versus inferiority (6 years to puberty), identity versus role confusion (adolescence), intimacy versus isolation (young adulthood), generativity versus stagnation (adulthood), and integrity versus despair (maturity). During intimacy versus isolation in the stage of young adulthood, there is psychosocial pressure for a person to form a close, stable interpersonal relationship. Thus, to the extent that one can attain an unconditional interchange and relationship, one will feel a sense of intimacy in feelings, ideas, and goals. If one cannot share and be shared, then one will feel a sense of isolation.

378. MICROBIOLOGY: ANSWER: B

(*Balows, 5/e, pp 323-330*) *Pneumocystis carinii, Mycobacterium tuberculosis*, and *Mycobacterium avium-intracellulare* are the respiratory pathogens most commonly seen in patients with AIDS. The diagnosis of these infections is best made from tissue. Minimally, the tissue sections should be stained with direct fluorescent antibody stains for *P. carinii* and an acid-fast stain for *Mycobacterium*. Methylene blue is a nonspecific stain. While the Gram stain (bacteria), lactophenol cotton blue stain (fungi), and Giemsa stain (parasites, multinucleated giant cells) are all useful, none would be the first choice for AIDS patients. A DNA probe based on the polymerase chain reaction (PCR) is available in research laboratories for *P. carinii* and *Mycobacterium* in sputum.

379. PATHOLOGY: ANSWER: E

(*Robbins, 4/e, pp 1416-1417*) Although several lesions within the brain may be associated with dystrophic or metaplastic calcification, the presence of a calcified tumorlike mass lesion in the cerebral hemispheres should arouse suspicion of oligodendroglioma. Oligodendrogliomas are often slow-growing gliomas composed of round cells with clear cytoplasm ("fried-egg appearance"); they generally occur in the fourth and fifth decades of life. However, some oligodendrogliomas do proliferate in a rapid and aggressive fashion and may be associated with a malignant astrocytoma component. The brown tumor associated with hypercalcemia of hyperparathyroidism is associated with osteitis fibrosa cystica of bone. Some metastatic carcinomas may show microcalcifications in the form of psammoma bodies, as do some meningiomas. Papillary carcinomas of the thyroid and ovary are the best examples of such lesions, but the calcifications found in papillary carcinomas are rarely of the degree and magnitude of those found in some oligodendrogliomas.

380. PHYSIOLOGY: ANSWER: C

381. PHYSIOLOGY: ANSWER: C

382. PHYSIOLOGY: ANSWER: D

(Guyton, 8/e, pp 304-307) The filtration fraction is the fraction of plasma filtered from the plasma flowing through the kidney, or GFR/RPF. Renal plasma flow is equal to the clearance of PAH; GFR is equal to the clearance of creatinine. Clearance of creatinine can be calculated using the following formula:

$$C_{cr} = \frac{U_{cr} \times \dot{V}}{P_{cr}}$$

$$= \frac{66 \times 2}{0.8}$$

$$= 165 \text{ mL / min}$$

Thus the filtration fraction = 165/750 = 0.22. At glucose concentrations below 150 to 200 mg/dL, the kidneys will reabsorb all the glucose passing through the kidney. The filtered load of glucose = $P_{glu} \times$ GFR = 120 mg/dL $\times$ 165 mL/min = 198 mg/min. Since all of this will be reabsorbed, the kidneys reabsorb approximately 200 mg of glucose per minute.

383. PATHOLOGY: ANSWER: C

(Robbins, 4/e, p 821) There are multiple causes for enlarged gingivae, some of which are physiologic and transient and require no investigative or therapeutic measures. Among these is pregnancy, which under the stimulation of hormones produces a vascular proliferation that presents histology similar to that of pyogenic granuloma. Another physiologic response is that seen at puberty. Nutritional disorders such as vitamin C deficiency can also lead to enlarged gums. Phenytoin (Dilantin) has been known to cause enlargement of the gums in some patients. If the setting does not suggest a physiologic response, consideration should be given to leukemia, especially monocytic leukemia, which can present enlarged gums as the initial manifestation. Thus, a complete blood count, including a Schilling differential to enumerate the white cells, is indicated.

384. PHYSIOLOGY: ANSWER: C

385. PHYSIOLOGY: ANSWER: D

(Berne, 2/e, pp 806-813. Rose, 3/e, pp 270-272, 469-473) The Henderson and Henderson-Hasselbalch equations for the bicarbonate buffer system are used to evaluate acid-base status. The Henderson equation is

$$\left[H^+\right] (\text{in nmol / L}) = 24 \times \frac{Pa_{CO_2} (\text{in mmHg})}{\left[HCO_3^-\right] (\text{in mmol / L})}$$

The conversion factors for the units and the K_a are "contained" in the number 24. The Henderson-Hasselbalch equation is the logarithmic form of this equation:

$$pH = 6.10 + \log \left(\frac{\left[HCO_3^-\right] \text{(in mmol / L)}}{0.03 \dfrac{\text{mmol / L}}{\text{mmHg}} \times Pa_{CO_2} \text{(in mmHg)}} \right)$$

The normal values are $[H^+] = 40$ nmol/L, pH = 7.40, $Pa_{CO_2} = 40$ mmHg, and $[HCO_3^-] = 24$ mmol/L. In practice it is easier to use the Henderson equation. For patient X the following would result:

$$50 \text{ nmol / L} = 24 \times \frac{30 \text{ mmHg}}{\left[HCO_3^-\right]}$$

$$\left[HCO_3^-\right] = 14.4 \text{ mmol / L}$$

The patient has a higher-than-normal concentration of H^+ ion (a lower-than-normal pH). This indicates that an acidosis is present. She also has a lower-than-normal Pa_{CO_2}, which indicates she is hyperventilating. Her bicarbonate level is also low. She most likely has a metabolic acidosis. Bicarbonate was consumed in the buffering of fixed acid. The Pa_{CO_2} is low because of stimulation of ventilation by the low arterial pH (respiratory compensation).

The following results can be obtained for this patient:

$$\left[H^+\right] = 24 \times \frac{30 \text{ mmHg}}{22 \text{ mmol / L}}$$

$$\left[H^+\right] = 33 \text{ nmol / L (pH = 7.49)}$$

The patient has a lower-than-normal concentration of H^+ ion (higher-than-normal pH) and therefore has an alkalosis. Since the Pa_{CO_2} is lower than normal, the patient is hyperventilating and the alkalosis is therefore a respiratory alkalosis. The bicarbonate concentration is only slightly below normal and this reduction is due to the decrease in Pa_{CO_2} (acute respiratory alkalosis). Over time, if the hyperventilation continues, the kidneys would excrete bicarbonate and further reduce the plasma bicarbonate level to raise the arterial $[H^+]$ back toward normal (chronic respiratory alkalosis).

386. PHARMACOLOGY: ANSWER: E

(DiPalma, 3/e, p 443) Cromolyn inhibits the release of mediators from mast cells, including histamine and slow reacting substance of anaphylaxis (SRS-A). This prevents allergically induced bronchospasm. Cromolyn is of no use in an acute asthmatic attack but is of considerable help in prophylaxis of asthmatic attacks, particularly in children.

387. MICROBIOLOGY: ANSWER: D

(Davis, 4/e, pp 379-383) The graph shown in the question exhibits hemagglutinating antibody responses to primary and secondary immunization with any standard antigen. Curve B represents the early response to primary immunization, which is chiefly an IgM response. Rechallenge elicits an accelerated response that mainly involves IgG and occurs 2 to 5 days after reimmunization. IgM has a molecular weight of 900,000 and is a pentamer that the fetus can produce quite early in gestation.

388. PATHOLOGY: ANSWER: C

(Richart, pp 1951-1959. Robbins, 4/e, pp 1135, 1142-1143) Cervical condylomata, particularly flat condylomata, although benign are considered to be precursors of cervical intraepithelial neoplasia (CIN), which comprises both dysplasia and carcinoma in situ (CIS). Histologically, these condylomata consist of connective tissue stroma covered by hyperplastic epithelium with prominent perinuclear cytoplasmic vacuolization (koilocytosis). Koilocytotic cells are characteristic of human papilloma virus (HPV) infection. More than 50 genotypes of HPV are known at present, and condylomata acuminata are associated with types 6/11 while HPV types 16/18 are usually present in CIN. Following an abnormal Pap smear report suggesting condyloma, CIN, or possible invasive carcinoma, workup of the patient should include colposcopy, multiple cervical punch biopsies, and endocervical curettage to distinguish patients who have invasive cancer, CIN, or flat condylomata.

389. ANATOMY: ANSWER: C

(Hollinshead, 4/e, pp 936, 949-950) The stapedius muscle is innervated by a branch of the facial nerve (CN VII) given off within the facial canal. Paralysis of this muscle results in loss of a reflex that dampens movement of the stapes—as a result, normal sounds are perceived as annoyingly loud. The tensor tympani muscle is innervated by a twig from the mandibular division of the trigeminal nerve (CN V3). The chorda tympani conveys taste from the anterior two-thirds of the tongue and secretomotor neurons for the submandibular and sublingual glands. The tympanic nerve of Jacobson, a branch of the glossopharyngeal nerve (CN IX), conveys secretomotor neurons to the parotid gland.

390. ANATOMY: ANSWER: C

(Hollinshead, 4/e, pp 936, 950-951) The greater superficial petrosal nerve leaves the facial nerve (CN VII) at the geniculate ganglion. It carries secretomotor neurons from the superior salivatory nucleus to the pterygopalatine ganglion, joining along the way with the sympathetic deep petrosal nerve to become the nerve of the pterygoid canal.

391. ANATOMY: ANSWER: C

(April, 2/e, pp 498-500) The facial nerve conveys special visceral efferent (branchiomotor) neurons to the muscles of facial expression and, in the facial canal, gives off a branch to the stapedius muscle. In addition, it carries general visceral efferent (secretomotor) neurons to the salivary glands and special visceral afferent (taste) fibers from the anterior two-thirds of the tongue via the chorda tympani. It also carries general visceral efferent (secretomotor) neurons to the lacrimal gland via the greater superficial petrosal nerve, and general somatic afferent neurons from the region of the external ear. Because lacrimation and balance are unaffected and the patient complains of hyperacusia, the lesion must be located in the facial canal between the geniculate ganglion and the origin of the nerve to the stapedius muscle. A lesion at the stylomastoid foramen or within the parotid gland would produce facial paralysis only. A lesion in the internal auditory meatus would affect lacrimation and might also involve the acousticovestibular nerve (CN VIII).

392. PHYSIOLOGY: ANSWER: D

(West, 12/e, pp 554-558) Under normal conditions the V/Q ratio in both lungs is the same, so that mixed alveolar gas in the trachea has the same P_{O_2} and P_{CO_2} as arterial blood (P_{O_2} = 100 mmHg, P_{CO_2} = 40 mmHg). Immediately following complete occlusion of one pulmonary artery, however, equal ventilation of both lungs continues, but all blood flow is directed to one lung. Equal volumes of gas will continue to mix in the trachea, but the gas from the occluded lung, which now represents alveolar dead space, will be unchanged from room air (P_{O_2} = 150 mmHg, P_{CO_2} = 0.3 mmHg); and gas from the functioning lung will still be normal (P_{O_2} = 100 mmHg, P_{CO_2} = 40 mmHg). Since equal volumes of gas mix, P_{O_2} in the trachea will be (150 + 100) ÷ 2, or 125 mmHg, and P_{CO_2} will be (40 + 0.3) ÷ 2, or 20 mmHg. Such values could occur in normal lungs following hyperventilation but would be accompanied by changes in arterial P_{CO_2}.

393. MICROBIOLOGY: ANSWER: B

(Davis, 4/e, pp 421-422) The Arthus reaction is a classic inflammatory response involving a cellular infiltrate provoked by antigen and antibody in much larger quantities than those required for passive cutaneous anaphylaxis. The edema of cutaneous anaphylaxis appears within 10 min and resolves within 30 min of antigen injection, but the polymorphonuclear leukocyte infiltrate of an Arthus reaction appears after more than an hour, peaks at 3 to 4 h, and resolves within 12 h. The severity of the Arthus reaction is proportional to the amount of antigen and antibody reacting. With high antigen-antibody concentrations, necrosis may result.

394. PHYSIOLOGY: ANSWER: B

(Ganong, 15/e, pp 510-516) The QRS complex represents the depolarization of the ventricles. Under normal circumstances, the ventricles are depolarized by the propagation of the action potential along the His-Purkinje system. Anything that interferes with the normal propagation pathway will cause the QRS complex to appear abnormal. In principle, the QRS complex can appear normal under all the listed situations if the common bundle branch is the first part of the ventricle to depolarize. However, this is unlikely to happen when the action potential does not pass along its normal path from the atria to the ventricles. The pathway is least likely to be interrupted in a first-degree heart block.

395. BIOCHEMISTRY: ANSWER: A

(Stryer, 3/e, pp 215-216) Carboxypeptidase is a protease that catalyzes the hydrolysis of a peptide bond immediately adjacent to a free α-carboxyl group. This property permits its use in the sequencing of a peptide from the *C*-terminal end. Pyruvate carboxylase catalyzes the addition of CO_2 to pyruvate to form oxaloacetate; isocitrate dehydrogenase converts isocitrate to α-ketoglutarate and CO_2; phosphoenolpyruvate carboxykinase catalyzes the conversion of oxaloacetate to phosphoenolpyruvate and CO_2.

396. ANATOMY: ANSWER: D

(Hollinshead, 4/e, p 506) Bronchopulmonary segments, the anatomic and functional units of the lung, are roughly pyramidal in shape, have apices directed toward the hilum of the lung, and are separated from each other by connective tissue septa. Each bronchopulmonary segment is supplied by one tertiary or segmental bronchus, along with a branch of the pulmonary artery. Although the segmental bronchus and artery tend to be centrally located, the veins do not accompany the arteries but tend to be located subpleurally and between bronchopulmonary segments. Indeed, at surgery the intersegmental veins are useful in defining intersegmental planes.

397. CELL BIOLOGY: ANSWER: B

(Alberts, 2/e, pp 288-291. Stevens, p 19. Widnell, pp 82-83) Spectrin is a cytoskeletal protein noncovalently bound to the cytosolic side of the membrane in red blood cells. It is found in high concentrations in red blood cell membranes. Actin readily forms complexes with spectrin, which may be considered as an intermediate filament protein. Spectrin is bound to another protein, ankyrin, which attaches to transmembrane band 3 and limits lateral diffusion. Binding of cell membranes to extracellular matrix molecules such as fibronectin and laminin through transmembrane proteins such as integrins (i.e., fibronectin and laminin receptors) also reduces lateral diffusion by tethering the membrane to the extracellular matrix. Talin is a protein that anchors actin, a major cytoskeletal element, to transmembrane proteins such as the integrins. The binding of an antibody to a cell surface antigen should result in patching and eventual capping on the extracellular surface as a result of ligand-receptor binding. This process involves an increase in lateral mobility of proteins in the lipid bilayer.

398. BIOCHEMISTRY: ANSWER: D

(Stryer, 3/e, pp 500-501) Carbamoyl phosphate is a precursor of urea. Glutamic acid and glutamine are important in the scavenging of ammonia. Only nitrate is not involved in keeping ammonia below toxic concentrations.

399. PHARMACOLOGY: ANSWER: E

(DiPalma, 3/e, pp 438-439) The most likely mechanism of methylxanthine action is by inhibition, not stimulation, of phosphodiesterase. This leads to an increase in cytosolic cyclic AMP and subsequent relaxation of smooth muscle. Methylxanthines also antagonize adenosine receptors. Since adenosine causes bronchoconstriction, this antagonism may also contribute to the bronchodilating action.

400. BEHAVIORAL SCIENCE: ANSWER: D

(Weiss, pp 31-34, 321-328) The incidence and prevalence of coronary heart disease caused by atherosclerosis have been linked to the type A pattern of coronary-prone behavior. Persons considered to be type A are more prone to respond to environmental challenges (social, psychological, or physical) with increased physiological responses. These behaviorally induced physiological responses over a lifetime appear to be linked to the development of certain cardiovascular disorders. Laboratory studies of humans have found that type A persons exhibit an increased level of systolic blood pressure, heart rate, plasma norepinephrine, plasma epinephrine, and cortisol and a decrease in occipital alpha activity. Other increases associated with coronary heart disease include those in serum cholesterol, serum triglycerides, platelet aggregation, clotting time, and serum corticotropin.

401. BIOCHEMISTRY: ANSWER: C

(Stryer, 3/e, pp 436-437) The reduction of oxidized glutathione is central to its role as a sulfhydryl "buffer." Its irreversible oxidation is not at all normal. Certain antimalarial drugs cause excessive oxidation of glutathione, which is then unavailable to protect against oxidation of proteins and to keep hemoglobin in the ferrous state. One result of this is anemia.

402. BIOCHEMISTRY: ANSWER: B

(Stryer, 3/e, pp 454-455) Skeletal muscle produces glucose-6-phosphate (via glucose-1- phosphate) from glycogen, but lacking glucose-6-phosphatase, cannot produce free glucose. The production of glucose from glycogen is restricted to the liver.

403. BIOCHEMISTRY: ANSWER: B

(Stryer, 3/e, pp 604, 614) The vitamin shown is folic acid. It is involved in the synthesis of ATP, GTP, and TTP, but not in the synthesis of CTP.

404. BIOCHEMISTRY: ANSWER: A

(Stryer, 3/e, pp 363, 373, 439) The conversion of phosphoenolpyruvate to pyruvate, the action of the enzyme pyruvate kinase, is essentially irreversible. In hepatic gluconeogenesis, the coupling of two other enzymes, pyruvate carboxylase and phosphoenolpyruvate carboxykinase, permit carbon atoms to bypass pyruvate kinase on the way from pyruvate to glucose.

405. PATHOLOGY: ANSWER: B.

(Robbins, 4/e, pp 51-52) The enzymatic defect that exists in chronic granulomatous disease of childhood does not impair chemotaxis or the cell's ability to engulf bacteria, but rather involves a failure to produce hydrogen peroxide after engulfment. Chemotactic defects resulting in inhibition of the capacity of leukocytes to infiltrate an area of infection or injury may be due to intracellular defects, as found in Chédiak-Higashi syndrome, in other genetic defects, and in diabetes mellitus. Chronic renal failure and other liver disease may be associated with factors in the circulation that impair chemotaxis. Whether newborn infants are full-term or immature, neonatal leukocytes have a temporary chemotactic defect that is corrected with increasing age.

406. PATHOLOGY: ANSWER: D

(Robbins, 4/e, pp 1298-1299, 1302-1306) Patients of either sex in the fourth to sixth decade who develop oral vesicles followed by disseminated bullae are likely to have pemphigus vulgaris, one of the blistering (bullous) dermatoses. The differential diagnosis in this setting is widespread and can include various forms of erythema multiforme (or Stevens-Johnson syndrome in the young) and bullous pemphigoid, as well as pemphigus vulgaris. Common to most bullous dermatoses is the presence of epidermal cell separation, which produces spaces and clefts (acantholysis) that are visible in ordinary tissue sections and specific to location within the epidermis. The bullae may be subcorneal, intraepidermal, suprabasal, or subepidermal, and multiple diseases can be grouped according to acantholysis location. To categorize the type of disease further, direct immunofluorescence testing can be done on a fresh skin lesion, using antibodies to immunoglobulins, fibrin, and complement. Pemphigus vulgaris shows a characteristic "basket-weave" pattern in the epidermis to IgG, IgA is found at the tips of the dermal papillae in dermatitis herpetiformis, and linear bands of IgG and complement are found in the subepidermal zones in bullous pemphigoid, whereas erythema multiforme has no immunofluorescent pattern. In DLE, direct immunofluorescence shows a granular band of immunoglobulin and complement at the dermoepidermal junction (lupus band test), and this may be present in "normal" skin in patients with systemic lupus erythematosus.

407. BIOCHEMISTRY: ANSWER: D

(Stryer, 3/e, pp 7-11) Peptide bonds are covalent in nature and are considered to be elements of the primary structure of a protein. Hydrogen bonds, electrostatic interaction, hydrophobic interactions, and van der Waals forces are all noncovalent and may contribute to the stabilization of secondary, tertiary, and, if present, quaternary structure. Quaternary structure is defined as the noncovalent interaction of two or more polypeptides.

408. GENETICS: ANSWER: E

(Gelehrter, pp 76-88, 102-106. Thompson, 5/e, pp 106-113) Sickle cell anemia is an autosomal recessive hemoglobinopathy that occurs in approximately 1 in 500 births in the black population. It is caused by a single nucleotide substitution in codon 6 of the hemoglobin gene. This mutation abolishes a restriction site for the enzyme MstII and thus can be detected by Southern blot analysis after digestion with this enzyme. It can similarly be detected when a PCR product is digested with this enzyme. ASO hybridization also detects single base pair substitutions. DNA sequencing will detect any change in the order of bases in a DNA fragment. Western blotting is a technique that examines the size and amount of mutant protein in cell extracts and therefore would not identify the sickle cell mutation in DNA.

409. PHARMACOLOGY: ANSWER: B

(DiPalma, 3/e, pp 382-383, 391-394) Like digitalis, verapamil is most useful in management of supraventricular arrhythmias, including atrial fibrillation, flutter, and atrial tachycardia especially caused by reentry mechanisms. Among its adverse reactions are constipation and headaches, although it also causes bradycardia, hypertension, and cardiac failure. Its main mechanism of action is on the slow-response calcium channels. The oral bioavailability of verapamil is only about 15 percent because of extensive first-pass biotransformation.

410. BEHAVIORAL SCIENCE: ANSWER: A

(Andres, pp 53-80) The elderly use a disproportionate amount of health care resources. With the estimated increase in the 65-and-over age group from 24 million (11 percent) to 38 million (14 percent), there will be an increased demand for health care services between now and the year 2000. Also, with the shift of health problems from acute illness to more chronic and debilitating conditions, there will be additional need to increase continuing and long-term health care. Data show women to be heavier users of services than men. With the increase in the elderly and with the female population progressively outnumbering the male population, the need for health care and help with social and economic problems in elderly females is expected to increase. The greater part of the sex differential in human longevity is considered to be the cumulative result of excessive male mortality throughout the life span. This is because of increased risk behaviors and habits that increase the vulnerability of males to health problems. The decline in mortality of the elderly over the past 15 years is partly due to the reduction in mortality from cardiovascular disease, improvements in medical science, and a shift in the population toward more information and concern with their own health behaviors.

411. BEHAVIORAL SCIENCE: ANSWER: B

(Andres, pp 53-71) The number of persons over the age of 65 has continued to increase in this century from 3.1 million in 1900 to 24.1 million in 1978. The proportion of the over-65 age group also rose from 4.1 percent to 11 percent. By the year 2000, the over-65 age group was expected to increase to 32 million and 12 percent of the population, but present declining mortality could result in 38 million and about 14 percent of the total population by the year 2000. In 1978, the over-75 age group was 38 percent of the total over-65 age group, and it is estimated that by 2003 the over-75 age group will have increased to about 47 percent. Thus more of the population has been reaching the 65-and-older age group, and living longer, and will continue to do so. Also, among the elderly the female population has progressively outnumbered the male population. The 80 men per 100 women in the 65-to-69-year-old group decreases to 45 men per 100 women in the 85-and-over population. While the 80 men per 100 women in the 65-to-69-year-old group is expected to rise slightly to 82 by the year 2000 and 83 by 2020, the men per 100 women in the 85-and-over group is expected to fall to 39 in 2000 and slightly below

39 by 2020. Only about 5 percent of the over-65 age group are in institutions (e.g., long-term facilities and nursing homes), which means that about 95 percent of the elderly are attempting to live with some measure of independence. A disproportionate number of the institutionalized elderly are white (94 percent), and there are twice as many females as males.

412. HISTOLOGY: ANSWER: B

(Junqueira, 7/e, pp 486-487. Newell, 7/e, pp 492-500. Stevens, p 201. Vaughan, 13/e, pp 203-206) The ophthalmoscope is extremely valuable in the analysis of vascular changes in diseases such as hypertension and diabetes mellitus. In diabetes mellitus one of the major complications is diabetic retinopathy. In diabetic retinopathy the pathologic changes usually begin with thickening of the basement membrane of the small retinal vessels. The abnormal vessels develop microaneurysms, which leak and hemorrhage with resultant ischemia of the retinal tissue. New vessels proliferate in response to the ischemia and the production of angiogenic factors. Loss of phagocytosis by the RPE occurs in retinal dystrophy but is not a characteristic of diabetic retinopathy.

413. BIOCHEMISTRY: ANSWER: D

(Stryer, 3/e, pp 482, 487-488) Citrate activates acetyl CoA carboxylase which produces malonyl CoA, the immediate source of most of the carbon in fatty acids. High levels of citrate in the cytoplasm inhibit glycolysis by an allosteric inhibition of phosphofructokinase-1. Liver mitochondria produce acetyl CoA by oxidizing fatty acids. In modest quantities, this acetyl CoA is fed into the Krebs cycle and is oxidized to CO_2. If the level of acetyl CoA exceeds the ability of liver mitochondria to oxidize it, ketone bodies are made and shipped via the bloodstream to other tissues which use them as fuel.

414. PHYSIOLOGY: ANSWER: E

(Davenport, 5/e, pp 159-165. Guyton, 8/e, pp 721-722 and 729) Primary bile acids are synthesized from cholesterol in the liver by the addition of hydroxyl and carboxyl groups and are secreted as amide conjugates with either taurine or glycine. Dehydroxylation of these compounds by intestinal bacteria forms secondary bile acids. Both primary and secondary bile acids are reabsorbed primarily by active transport in the ileum and return to the liver via the portal vein, where they are reutilized and secreted in the bile. Because they possess both hydrophobic and hydrophilic properties, bile acids and their salts accumulate at lipid-water interfaces and thus emulsify dietary fat and promote its hydrolysis. Sulfation of bile acids occurs in the liver. Since sulfated bile acids are not reabsorbed in the ileum, this process is a major route for excretion of these compounds in the feces.

415. BIOCHEMISTRY: ANSWER: A

(Stryer, pp 161-162) The synthesis of 2,3-DPG occurs as a side reaction in glycolysis. This compound lowers the affinity of hemoglobin for oxygen, which facilitates the delivery of oxygen to the tissues. Its synthesis bypasses one of the ATP-yielding steps of glycolysis, so the net ATP yield from a molecule of glucose is zero. For this reason, the red blood cell can never convert all of its 1,3-DPG into 2,3-DPG, since glycolysis is its only source of energy.

416. PATHOLOGY: ANSWER: E

(Robbins, 4/e, pp 1389-1390, 1409-1411) The clinical constellation of altered sensorium and papilledema should call to mind the presence of intracranial pressure, regardless of the cause, which can be due to cerebral edema, tumor mass, or, more commonly, intracranial bleeding with hematoma formation. If the pressure is severe enough, downward displacement of the cerebellar tonsils into the foramen magnum may occur, producing further compression on the brainstem with consequent hemorrhage into the pons and midbrain (Duret hemorrhages). This is nearly always associated with death, since the vital centers, including respiratory control, are located in these regions. Subdural as well as epidural hemorrhages are sufficient to cause critical downward displacement of the cerebellar tonsils. The situation can be remedied with appropriate neurosurgical intervention. In this situation, the downward displacement could be due to hemorrhage-hematoma formation into the posterior intracranial fossa, caused by either a direct (coup) or an indirect (contracoup) blow to the occiput.

417. MICROBIOLOGY: ANSWER: C

(Wilson, 12/e, pp 689-692) Chronic mononucleosis is an identifiable syndrome characterized by chronic malaise, often following an emotional stress. Unfortunately, chronic mononucleosis has become a catch-all diagnosis for many other infectious and noninfectious diseases that cause similar symptoms. The EBV pattern usually seen in chronic infectious mononucleosis is characterized by an elevated VCA-IgG, no VCA-IgM, low-titer EBNA, and elevated EA (1:20).

418. PHYSIOLOGY: ANSWER: B

(Ganong, 15/e, pp 589-590. Guyton, 8/e, pp 264-268) Reduction in blood volume by hemorrhage decreases venous return. The arterial baroreceptors are stretched to a lesser degree and sympathetic outflow is increased. There is reflex tachycardia and generalized vasoconstriction, except in the blood vessels of the brain and heart. Venoconstriction helps to maintain the filling pressure of the heart.

419. PHARMACOLOGY: ANSWER: A

(DiPalma, 3/e, pp 561-562) The potential serious adverse effect of bleomycin is pneumonitis and pulmonary fibrosis. This adverse effect appears to be both age- and dose-related. The clinical onset is characterized by decreasing pulmonary function, fine rales, cough, and diffuse basilar infiltrates. This complication develops in approximately 5 to 10 percent of patients treated with bleomycin. Thus, extreme caution must be used in patients with a preexisting history of pulmonary disease. All the other drugs listed in the question are effective against carcinomas and have not been associated with significant lung toxicity.

420. BIOCHEMISTRY: ANSWER: B

(Stryer, 3/e, pp 703-710) An important difference between RNA synthesis and DNA synthesis is that the former does not need a primer, but starts with pppG or pppA.

Bibliography

Bibliography

Alberts B, Bray D, Lewis J, et al: *Molecular Biology of the Cell,* 2/e. New York, Garland, 1989.

AMA Drug Evaluations Annual 1991, 7/e. Chicago, American Medical Association, 1986.

Anderson WA, Kissane JM (eds): *Pathology,* 9/e. St. Louis, CV Mosby, 1989.

April EW: *Anatomy,* 2/e. New York, John Wiley & Sons, 1990.

Ash LR, Orihel TC: *Atlas of Human Parasitology,* 3/e. Chicago, ASCP, 1990.

Balows A, et al: *Manual of Clinical Microbiology,* 5/e. Washington, DC, American Society for Microbiology, 1991.

Balows A, Hausler WJ, Lennette EH: *Laboratory Diagnosis of Infectious Diseases: Principles and Practice,* vols. 1 and 2. New York, Springer-Verlag, 1988.

Baron S: *Medical Mircobiology,* 3/e. New York, Churchill Livingstone, 1991.

Bell RR: *Marriage and Family Interaction,* 6/e. Homewood, IL, Dorsey Press, 1983.

Berne RM, Levy MN: *Physiology,* 2/e. St. Louis, CV Mosby, 1988.

Braunwald E, et al: *Harrison's Principles of Internal Medicine,* 11/e. New York, McGraw-Hill, 1987.

Brenner C: *An Elementary Textbook of Psychoanalysis.* New York, Doubleday, 1974.

Brown HW, Neva FA: *Basic Clinical Parasitology,* 5/e. East Norwalk, CT, Appleton & Lange, 1983.

Coe FL, Favus MJ (eds): *Disorders of Bone and Mineral Metabolism.* New York, Raven, 1992.

Conger JJ, Petersen AC: *Adolescence and Youth: Psychological Development in a Changing World,* 3/e. New York, Harper & Row, 1983.

Cotran RS, Kumar V, Robbins SL: *Robbins Pathologic Basis of Disease,* 4/e. Philadelphia, WB Saunders, 1989.

Counte MA, Christman LP: *Interpersonal Behavior and Health Care.* Boulder, CO, Westview Press, 1981.

Davenport HW: *Physiology of the Digestive Tract,* 5/e. Chicago, Year Book Medical Publishers, 1982.

Davis BD, et al: *Microbiology,* 4/e. New York, Harper & Row, 1990.

DiPalma JR, DiGregorio GJ: *Basic Pharmacology in Medicine,* 3/e. New York, McGraw-Hill, 1990.

Erlandsen SL, Magney JE: *Color Atlas of Histology.* St. Louis, CV Mosby, 1992.

Fitzpatrick TB, et al: *Color Atlas and Synopsis of Clinical Dermatology,* 2/e. New York, McGraw-Hill, 1992.

Ganong WF: *Review of Medical Physiology,* 14/e. Norwalk, CT, Appleton & Lange, 1989.

Ganong WF: *Review of Medical Physiology,* 15/e. East Norwalk, CT, Appleton & Lange, 1991.

Gelehrter TD, Collins FS: *Principles of Medical Genetics.* Baltimore, Williams & Wilkins, 1990.

Gilman AG, et al (eds): *The Pharmacological Basis of Therapeutics,* 8/e. New York, Macmillan, 1990.

Goldstein A: *Biostatistics: An Introductory Text.* New York, Macmillan, 1964.

Guyton, AC: *Textbook of Medical Physiology,* 8/e. Philadelphia, WB Saunders, 1991.

Hollinshead WH, Rosse C: *Textbook of Anatomy,* 4/e. New York, Harper & Row, 1985.

Howard BJ, Klass J, Rubin SJ, Weissfeld AS, Tilton RC: *Clinical and Pathogenic Microbiology.* St. Louis, CV Mosby, 1987.

Jawetz E, Melnick JL, Adelbert EA: *Review of Medical Microbiology,* 19/e. East Norwalk, CT, Appleton & Lange, 1991.

Junqueira LC, Carneiro J, Kelley RO: *Basic Histology,* 7/e. East Norwalk, CT, Appleton & Lange, 1992.

Kandel ER, Schwartz JH, Jessell TM: *Principles of Neural Science,* 3/e. New York, Elsevier, 1991.

Katzung BG: *Basic and Clinical Pharmacology,* 4/e. East Norwalk, CT, Appleton & Lange, 1989.

Kelly DE, Wood RL, Enders AC: *Bailey's Textbook of Microscopic Anatomy,* 18/e. Baltimore, Williams & Wilkins, 1984.

Lerner RM, Galambos NL: *Experiencing Adolescence: A Sourcebook for Parents, Teachers, and Teens.* New York, Garland, 1984.

Mandell GL, Douglas RG, Bennett JE: *Principles and Practice of Infectious Disease,* 3/e. New York, John Wiley & Sons, 1990.

Moore KL: *The Developing Human: Clinically Oriented Embryology,* 4/e. Philadelphia, WB Saunders, 1988.

Mussen PH, Conger JJ, Kagan J, Huston AC: *Child Development and Personality,* 6/e. New York, Harper & Row, 1984.

Newell FW: *Ophthalmology: Principles and Concepts,* 7/e. St. Louis, CV Mosby, 1992.

Noback CR, Demarest RJ: *The Human Nervous System: Basic Principles of Neurobiology,* 3/e. New York, McGraw-Hill, 1981.

Richart RM: Causes and management of cervical intraepithelial neoplasia. *Cancer* 60:1951-1959, 1987.

Robbins SL, Cotran RS: *Pathologic Basics of Disease,* 4/e. Philadelphia, WB Saunders, 1989.

Roitt I, Brostoff J, Male D: *Immunology,* 2/e. St. Louis, CV Mosby, 1989.

Rose BD: *Clinical Physiology of Acid-Base and Electrolyte Disorders,* 3/e. New York, McGraw-Hill, 1989.

Rose NR, Bigazzi PE: *Methods in Immunodiagnosis,* 3/e. New York, John Wiley & Sons, 1987.

Ross MH, Reith EJ, Romrell LJ: *Histology: A Text Atlas,* 2/e. Baltimore, Williams & Wilkins, 1989.

Schuster CS, Ashburn SS: *The Process of Human Development: A Holistic Approach,* 2/e. Boston, Little, Brown, 1986.

Schwarz WB, Wolfe HJ, Pauker SG: Pathology and probabilities: A new approach to interpreting and reporting biopsies. *N Engl J Med* 305:917, 1981.

Simons RC (ed): *Understanding Human Behavior in Health and Illness,* 3/e. Baltimore, Williams & Wilkins, 1985.

Stevens A, Lowe J: *Histology.* New York, Gower Medical, 1992.

Stryer L: *Biochemistry,* 3/e. San Francisco, WH Freeman, 1988.

Suinn RM: *Fundamentals of Abnormal Psychology.* Chicago, Nelson-Hall, 1984.

Thompson MW, McInnes RR, Willard HF: *Genetics in Medicine,* 5/e. Philadelphia, WB Saunders, 1991.

Uitto J, Perejda AJ (eds): *Connective Tissue Disease: Molecular Pathology of the Extracellular Matrix.* New York, Dekker, 1987.

Vaughn D, Taylor A: *General Ophthalmology,* 13/e. East Norwalk, CT, Appleton & Lange, 1992.

Weiss SM, Herd JA, Fox BH (eds): *Perspectives on Behavioral Medicine,* vol 1. New York, Academic Press, 1981.

West JB: *Best & Taylor's Physiological Basis of Medical Practice,* 11/e. Baltimore, Williams & Wilkins, 1985.

West JB: *Best & Taylor's Physiological Basis of Medical Practice,* 12/e. Baltimore, Williams & Wilkins, 1990.

Wheater PR, Burkitt HG, Daniels VG: *Functional Histology: A Text and Colour Atlas,* 2/e. New York, Churchill Livingstone, 1987.

Widnell CC, Pfenninger KH: *Essential Cell Biology.* Baltimore, Williams & Wilkins, 1990.

Williams SJ, Torrens PR: *Introduction to Health Services,* 3/e. New York, John Wiley & Sons, 1988.

Wilson JD, Braunwald E, Isselbacher KJ, et al (eds): *Harrison's Principles of Internal Medicine,* 12/e. New York, McGraw-Hill, 1991.

Yen SSC, Jaffe RB: *Reproductive Endocrinology,* 3/e. Philadelphia, WB Saunders, 1991.

Index

Index

FOR BEST RESULTS WHEN RETURNING YOUR ANSWERS SHEETS TO BE COMPUTER EVALUATED AND GRADED

- *Carefully* tear along the perforated lines to remove the answer sheets.

- Keep your answers sheets flat and unmarked

- Enter your name and mailing address clearly and accurately. If you need more room than is available, print your name and address clearly on a Post-it™ note or a separate sheet of paper and include it with your answer sheets.

- Be sure to enter your four-digit identification code in the appropriate boxes of every page of the answer sheets.

- Fold the sheets carefully at the "fold" marks indicated, place them in a legal-size envelope, and mail them to:

  McGraw-Hill/PreTest
  HPD 28
  1221 Avenue of the Americas
  New York, NY 10020
  Attn: Test Marking.

- You will receive computer-evluated test results in the mail, but your answer sheets will not be returned

ISBN 0-07-052020-8

9 780070 520202   90000>

PreTest® Series

INSTRUCTIONS FOR MARKING THE ANSWER SHEETS

* Use a No. 2 Pencil.

* Make a dark mark to completely fill in the box in the appropriate grid position; your mark should fill the space within the box but not stray outside its lines; do not use slashes.

* If you erase, erase the change cleanly and completely.

* Print your Last Name, First Initial, Street Address, and Zip Code in the appropriate boxes and mark the corresponding grid boxes. This information MUST BE accurate since an optical scanner will generate the mailing label to return your score to the address as coded.

* Print your City and State in the appropriate boxes. Although the optical scanner generates this information for your mailing label from your Zip Code, the printed text can help us locate you in case of error.

* Enter the RED five-digit identification code (stamped on the demographics sheet) in the box in the upper left-hand corner of EACH answer sheet. Enter the numbers accurately; mark all corresponding grid boxes – even zeros.

* For each question on this examination, fill in the appropriate grid boxes on your answer sheet corresponding to it.

* If your answer sheets are not filled in according to these guidelines, the optical scanner will not read your answers correctly. Your sheets will be returned to you either unscored or with a message indicating that, due to incomplete erasures or faulty recording of answers, there may be inaccuracies in your score.

* When you complete the examination, (1) fold your demographics sheet and answer sheets carefully along the "fold here" lines, place them in an envelope, stamp with extra postage, and mail them to McGraw-Hill/PreTest, HPD 28, 1221 Avenue of the Americas, New York, NY 10020, Att: Test Marking; or (2) give the demographics sheet and answer sheet(s) to your proctor for mailing.

* McGraw-Hill/PreTest® determines the score for this test on the basis of the number of questions answered correctly, as is done by the national licensing board. Scores also show your standing among those others who have taken this version of this examination.

Published by McGraw-Hill, Inc. PreTest® Series.
Copyright © 1995 by McGraw-Hill, Inc.
All Rights Reserved. Printed in the U.S.A.

- Use a No. 2 Pencil, make dark marks, and erase changes completely
- Mark the five-digit code box on each sheet – this is VERY IMPORTANT
- Print your Last Name, First Initial, Street Address, and Zip Code in the top column of boxes and mark the corresponding box for each letter or number required
- Print your City and State in the appropriate boxes
- Be certain that your Zip Code is correct to ensure proper mailing of your test results

LAST NAME

⊏A⊐	⊏A⊐	⊏A⊐	⊏A⊐	⊏A⊐	⊏A⊐	⊏A⊐	⊏A⊐	⊏A⊐	⊏A⊐	⊏A⊐
⊏B⊐	⊏B⊐	⊏B⊐	⊏B⊐	⊏B⊐	⊏B⊐	⊏B⊐	⊏B⊐	⊏B⊐	⊏B⊐	⊏B⊐
⊏C⊐	⊏C⊐	⊏C⊐	⊏C⊐	⊏C⊐	⊏C⊐	⊏C⊐	⊏C⊐	⊏C⊐	⊏C⊐	⊏C⊐
⊏D⊐	⊏D⊐	⊏D⊐	⊏D⊐	⊏D⊐	⊏D⊐	⊏D⊐	⊏D⊐	⊏D⊐	⊏D⊐	⊏D⊐
⊏E⊐	⊏E⊐	⊏E⊐	⊏E⊐	⊏E⊐	⊏E⊐	⊏E⊐	⊏E⊐	⊏E⊐	⊏E⊐	⊏E⊐
⊏F⊐	⊏F⊐	⊏F⊐	⊏F⊐	⊏F⊐	⊏F⊐	⊏F⊐	⊏F⊐	⊏F⊐	⊏F⊐	⊏F⊐
⊏G⊐	⊏G⊐	⊏G⊐	⊏G⊐	⊏G⊐	⊏G⊐	⊏G⊐	⊏G⊐	⊏G⊐	⊏G⊐	⊏G⊐
⊏H⊐	⊏H⊐	⊏H⊐	⊏H⊐	⊏H⊐	⊏H⊐	⊏H⊐	⊏H⊐	⊏H⊐	⊏H⊐	⊏H⊐
⊏I⊐	⊏I⊐	⊏I⊐	⊏I⊐	⊏I⊐	⊏I⊐	⊏I⊐	⊏I⊐	⊏I⊐	⊏I⊐	⊏I⊐
⊏J⊐	⊏J⊐	⊏J⊐	⊏J⊐	⊏J⊐	⊏J⊐	⊏J⊐	⊏J⊐	⊏J⊐	⊏J⊐	⊏J⊐
⊏K⊐	⊏K⊐	⊏K⊐	⊏K⊐	⊏K⊐	⊏K⊐	⊏K⊐	⊏K⊐	⊏K⊐	⊏K⊐	⊏K⊐
⊏L⊐	⊏L⊐	⊏L⊐	⊏L⊐	⊏L⊐	⊏L⊐	⊏L⊐	⊏L⊐	⊏L⊐	⊏L⊐	⊏L⊐
⊏M⊐	⊏M⊐	⊏M⊐	⊏M⊐	⊏M⊐	⊏M⊐	⊏M⊐	⊏M⊐	⊏M⊐	⊏M⊐	⊏M⊐
⊏N⊐	⊏N⊐	⊏N⊐	⊏N⊐	⊏N⊐	⊏N⊐	⊏N⊐	⊏N⊐	⊏N⊐	⊏N⊐	⊏N⊐
⊏O⊐	⊏O⊐	⊏O⊐	⊏O⊐	⊏O⊐	⊏O⊐	⊏O⊐	⊏O⊐	⊏O⊐	⊏O⊐	⊏O⊐
⊏P⊐	⊏P⊐	⊏P⊐	⊏P⊐	⊏P⊐	⊏P⊐	⊏P⊐	⊏P⊐	⊏P⊐	⊏P⊐	⊏P⊐
⊏Q⊐	⊏Q⊐	⊏Q⊐	⊏Q⊐	⊏Q⊐	⊏Q⊐	⊏Q⊐	⊏Q⊐	⊏Q⊐	⊏Q⊐	⊏Q⊐
⊏R⊐	⊏R⊐	⊏R⊐	⊏R⊐	⊏R⊐	⊏R⊐	⊏R⊐	⊏R⊐	⊏R⊐	⊏R⊐	⊏R⊐
⊏S⊐	⊏S⊐	⊏S⊐	⊏S⊐	⊏S⊐	⊏S⊐	⊏S⊐	⊏S⊐	⊏S⊐	⊏S⊐	⊏S⊐
⊏T⊐	⊏T⊐	⊏T⊐	⊏T⊐	⊏T⊐	⊏T⊐	⊏T⊐	⊏T⊐	⊏T⊐	⊏T⊐	⊏T⊐
⊏U⊐	⊏U⊐	⊏U⊐	⊏U⊐	⊏U⊐	⊏U⊐	⊏U⊐	⊏U⊐	⊏U⊐	⊏U⊐	⊏U⊐
⊏V⊐	⊏V⊐	⊏V⊐	⊏V⊐	⊏V⊐	⊏V⊐	⊏V⊐	⊏V⊐	⊏V⊐	⊏V⊐	⊏V⊐
⊏W⊐	⊏W⊐	⊏W⊐	⊏W⊐	⊏W⊐	⊏W⊐	⊏W⊐	⊏W⊐	⊏W⊐	⊏W⊐	⊏W⊐
⊏X⊐	⊏X⊐	⊏X⊐	⊏X⊐	⊏X⊐	⊏X⊐	⊏X⊐	⊏X⊐	⊏X⊐	⊏X⊐	⊏X⊐
⊏Y⊐	⊏Y⊐	⊏Y⊐	⊏Y⊐	⊏Y⊐	⊏Y⊐	⊏Y⊐	⊏Y⊐	⊏Y⊐	⊏Y⊐	⊏Y⊐
⊏Z⊐	⊏Z⊐	⊏Z⊐	⊏Z⊐	⊏Z⊐	⊏Z⊐	⊏Z⊐	⊏Z⊐	⊏Z⊐	⊏Z⊐	⊏Z⊐

CODE

⊏0⊐	⊏0⊐	⊏0⊐	⊏0⊐	⊏0⊐
⊏1⊐	⊏1⊐	⊏1⊐	⊏1⊐	⊏1⊐
⊏2⊐	⊏2⊐	⊏2⊐	⊏2⊐	⊏2⊐
⊏3⊐	⊏3⊐	⊏3⊐	⊏3⊐	⊏3⊐
⊏4⊐	⊏4⊐	⊏4⊐	⊏4⊐	⊏4⊐
⊏5⊐	⊏5⊐	⊏5⊐	⊏5⊐	⊏5⊐
⊏6⊐	⊏6⊐	⊏6⊐	⊏6⊐	⊏6⊐
⊏7⊐	⊏7⊐	⊏7⊐	⊏7⊐	⊏7⊐
⊏8⊐	⊏8⊐	⊏8⊐	⊏8⊐	⊏8⊐
⊏9⊐	⊏9⊐	⊏9⊐	⊏9⊐	⊏9⊐

FIVE-DIGIT CODE
15833

Mark the five-digit code printed above in the code box here and on all other sides.

PLEASE FOLD HERE

FIRST INITIAL

⊏A⊐ ⊏B⊐ ⊏C⊐ ⊏D⊐ ⊏E⊐ ⊏F⊐ ⊏G⊐ ⊏H⊐ ⊏I⊐ ⊏J⊐
⊏K⊐ ⊏L⊐ ⊏M⊐ ⊏N⊐ ⊏O⊐ ⊏P⊐ ⊏Q⊐ ⊏R⊐ ⊏S⊐ ⊏T⊐
⊏U⊐ ⊏V⊐ ⊏W⊐ ⊏X⊐ ⊏Y⊐ ⊏Z⊐

CITY

STATE

STREET ADDRESS OR P.O. BOX

⊏A⊐	⊏A⊐	⊏A⊐	⊏A⊐	⊏A⊐	⊏A⊐	⊏A⊐	⊏A⊐	⊏A⊐	⊏A⊐	⊏A⊐	⊏A⊐	⊏A⊐	⊏A⊐	⊏A⊐	⊏A⊐
⊏B⊐	⊏B⊐	⊏B⊐	⊏B⊐	⊏B⊐	⊏B⊐	⊏B⊐	⊏B⊐	⊏B⊐	⊏B⊐	⊏B⊐	⊏B⊐	⊏B⊐	⊏B⊐	⊏B⊐	⊏B⊐
⊏C⊐	⊏C⊐	⊏C⊐	⊏C⊐	⊏C⊐	⊏C⊐	⊏C⊐	⊏C⊐	⊏C⊐	⊏C⊐	⊏C⊐	⊏C⊐	⊏C⊐	⊏C⊐	⊏C⊐	⊏C⊐
⊏D⊐	⊏D⊐	⊏D⊐	⊏D⊐	⊏D⊐	⊏D⊐	⊏D⊐	⊏D⊐	⊏D⊐	⊏D⊐	⊏D⊐	⊏D⊐	⊏D⊐	⊏D⊐	⊏D⊐	⊏D⊐
⊏E⊐	⊏E⊐	⊏E⊐	⊏E⊐	⊏E⊐	⊏E⊐	⊏E⊐	⊏E⊐	⊏E⊐	⊏E⊐	⊏E⊐	⊏E⊐	⊏E⊐	⊏E⊐	⊏E⊐	⊏E⊐
⊏F⊐	⊏F⊐	⊏F⊐	⊏F⊐	⊏F⊐	⊏F⊐	⊏F⊐	⊏F⊐	⊏F⊐	⊏F⊐	⊏F⊐	⊏F⊐	⊏F⊐	⊏F⊐	⊏F⊐	⊏F⊐
⊏G⊐	⊏G⊐	⊏G⊐	⊏G⊐	⊏G⊐	⊏G⊐	⊏G⊐	⊏G⊐	⊏G⊐	⊏G⊐	⊏G⊐	⊏G⊐	⊏G⊐	⊏G⊐	⊏G⊐	⊏G⊐
⊏H⊐	⊏H⊐	⊏H⊐	⊏H⊐	⊏H⊐	⊏H⊐	⊏H⊐	⊏H⊐	⊏H⊐	⊏H⊐	⊏H⊐	⊏H⊐	⊏H⊐	⊏H⊐	⊏H⊐	⊏H⊐
⊏I⊐	⊏I⊐	⊏I⊐	⊏I⊐	⊏I⊐	⊏I⊐	⊏I⊐	⊏I⊐	⊏I⊐	⊏I⊐	⊏I⊐	⊏I⊐	⊏I⊐	⊏I⊐	⊏I⊐	⊏I⊐
⊏J⊐	⊏J⊐	⊏J⊐	⊏J⊐	⊏J⊐	⊏J⊐	⊏J⊐	⊏J⊐	⊏J⊐	⊏J⊐	⊏J⊐	⊏J⊐	⊏J⊐	⊏J⊐	⊏J⊐	⊏J⊐
⊏K⊐	⊏K⊐	⊏K⊐	⊏K⊐	⊏K⊐	⊏K⊐	⊏K⊐	⊏K⊐	⊏K⊐	⊏K⊐	⊏K⊐	⊏K⊐	⊏K⊐	⊏K⊐	⊏K⊐	⊏K⊐
⊏L⊐	⊏L⊐	⊏L⊐	⊏L⊐	⊏L⊐	⊏L⊐	⊏L⊐	⊏L⊐	⊏L⊐	⊏L⊐	⊏L⊐	⊏L⊐	⊏L⊐	⊏L⊐	⊏L⊐	⊏L⊐
⊏M⊐	⊏M⊐	⊏M⊐	⊏M⊐	⊏M⊐	⊏M⊐	⊏M⊐	⊏M⊐	⊏M⊐	⊏M⊐	⊏M⊐	⊏M⊐	⊏M⊐	⊏M⊐	⊏M⊐	⊏M⊐
⊏N⊐	⊏N⊐	⊏N⊐	⊏N⊐	⊏N⊐	⊏N⊐	⊏N⊐	⊏N⊐	⊏N⊐	⊏N⊐	⊏N⊐	⊏N⊐	⊏N⊐	⊏N⊐	⊏N⊐	⊏N⊐
⊏O⊐	⊏O⊐	⊏O⊐	⊏O⊐	⊏O⊐	⊏O⊐	⊏O⊐	⊏O⊐	⊏O⊐	⊏O⊐	⊏O⊐	⊏O⊐	⊏O⊐	⊏O⊐	⊏O⊐	⊏O⊐
⊏P⊐	⊏P⊐	⊏P⊐	⊏P⊐	⊏P⊐	⊏P⊐	⊏P⊐	⊏P⊐	⊏P⊐	⊏P⊐	⊏P⊐	⊏P⊐	⊏P⊐	⊏P⊐	⊏P⊐	⊏P⊐
⊏Q⊐	⊏Q⊐	⊏Q⊐	⊏Q⊐	⊏Q⊐	⊏Q⊐	⊏Q⊐	⊏Q⊐	⊏Q⊐	⊏Q⊐	⊏Q⊐	⊏Q⊐	⊏Q⊐	⊏Q⊐	⊏Q⊐	⊏Q⊐
⊏R⊐	⊏R⊐	⊏R⊐	⊏R⊐	⊏R⊐	⊏R⊐	⊏R⊐	⊏R⊐	⊏R⊐	⊏R⊐	⊏R⊐	⊏R⊐	⊏R⊐	⊏R⊐	⊏R⊐	⊏R⊐
⊏S⊐	⊏S⊐	⊏S⊐	⊏S⊐	⊏S⊐	⊏S⊐	⊏S⊐	⊏S⊐	⊏S⊐	⊏S⊐	⊏S⊐	⊏S⊐	⊏S⊐	⊏S⊐	⊏S⊐	⊏S⊐
⊏T⊐	⊏T⊐	⊏T⊐	⊏T⊐	⊏T⊐	⊏T⊐	⊏T⊐	⊏T⊐	⊏T⊐	⊏T⊐	⊏T⊐	⊏T⊐	⊏T⊐	⊏T⊐	⊏T⊐	⊏T⊐
⊏U⊐	⊏U⊐	⊏U⊐	⊏U⊐	⊏U⊐	⊏U⊐	⊏U⊐	⊏U⊐	⊏U⊐	⊏U⊐	⊏U⊐	⊏U⊐	⊏U⊐	⊏U⊐	⊏U⊐	⊏U⊐
⊏V⊐	⊏V⊐	⊏V⊐	⊏V⊐	⊏V⊐	⊏V⊐	⊏V⊐	⊏V⊐	⊏V⊐	⊏V⊐	⊏V⊐	⊏V⊐	⊏V⊐	⊏V⊐	⊏V⊐	⊏V⊐
⊏W⊐	⊏W⊐	⊏W⊐	⊏W⊐	⊏W⊐	⊏W⊐	⊏W⊐	⊏W⊐	⊏W⊐	⊏W⊐	⊏W⊐	⊏W⊐	⊏W⊐	⊏W⊐	⊏W⊐	⊏W⊐
⊏X⊐	⊏X⊐	⊏X⊐	⊏X⊐	⊏X⊐	⊏X⊐	⊏X⊐	⊏X⊐	⊏X⊐	⊏X⊐	⊏X⊐	⊏X⊐	⊏X⊐	⊏X⊐	⊏X⊐	⊏X⊐
⊏Y⊐	⊏Y⊐	⊏Y⊐	⊏Y⊐	⊏Y⊐	⊏Y⊐	⊏Y⊐	⊏Y⊐	⊏Y⊐	⊏Y⊐	⊏Y⊐	⊏Y⊐	⊏Y⊐	⊏Y⊐	⊏Y⊐	⊏Y⊐
⊏Z⊐	⊏Z⊐	⊏Z⊐	⊏Z⊐	⊏Z⊐	⊏Z⊐	⊏Z⊐	⊏Z⊐	⊏Z⊐	⊏Z⊐	⊏Z⊐	⊏Z⊐	⊏Z⊐	⊏Z⊐	⊏Z⊐	⊏Z⊐
⊏0⊐	⊏0⊐	⊏0⊐	⊏0⊐	⊏0⊐	⊏0⊐	⊏0⊐	⊏0⊐	⊏0⊐	⊏0⊐	⊏0⊐	⊏0⊐	⊏0⊐	⊏0⊐	⊏0⊐	⊏0⊐
⊏1⊐	⊏1⊐	⊏1⊐	⊏1⊐	⊏1⊐	⊏1⊐	⊏1⊐	⊏1⊐	⊏1⊐	⊏1⊐	⊏1⊐	⊏1⊐	⊏1⊐	⊏1⊐	⊏1⊐	⊏1⊐
⊏2⊐	⊏2⊐	⊏2⊐	⊏2⊐	⊏2⊐	⊏2⊐	⊏2⊐	⊏2⊐	⊏2⊐	⊏2⊐	⊏2⊐	⊏2⊐	⊏2⊐	⊏2⊐	⊏2⊐	⊏2⊐
⊏3⊐	⊏3⊐	⊏3⊐	⊏3⊐	⊏3⊐	⊏3⊐	⊏3⊐	⊏3⊐	⊏3⊐	⊏3⊐	⊏3⊐	⊏3⊐	⊏3⊐	⊏3⊐	⊏3⊐	⊏3⊐
⊏4⊐	⊏4⊐	⊏4⊐	⊏4⊐	⊏4⊐	⊏4⊐	⊏4⊐	⊏4⊐	⊏4⊐	⊏4⊐	⊏4⊐	⊏4⊐	⊏4⊐	⊏4⊐	⊏4⊐	⊏4⊐
⊏5⊐	⊏5⊐	⊏5⊐	⊏5⊐	⊏5⊐	⊏5⊐	⊏5⊐	⊏5⊐	⊏5⊐	⊏5⊐	⊏5⊐	⊏5⊐	⊏5⊐	⊏5⊐	⊏5⊐	⊏5⊐
⊏6⊐	⊏6⊐	⊏6⊐	⊏6⊐	⊏6⊐	⊏6⊐	⊏6⊐	⊏6⊐	⊏6⊐	⊏6⊐	⊏6⊐	⊏6⊐	⊏6⊐	⊏6⊐	⊏6⊐	⊏6⊐
⊏7⊐	⊏7⊐	⊏7⊐	⊏7⊐	⊏7⊐	⊏7⊐	⊏7⊐	⊏7⊐	⊏7⊐	⊏7⊐	⊏7⊐	⊏7⊐	⊏7⊐	⊏7⊐	⊏7⊐	⊏7⊐
⊏8⊐	⊏8⊐	⊏8⊐	⊏8⊐	⊏8⊐	⊏8⊐	⊏8⊐	⊏8⊐	⊏8⊐	⊏8⊐	⊏8⊐	⊏8⊐	⊏8⊐	⊏8⊐	⊏8⊐	⊏8⊐
⊏9⊐	⊏9⊐	⊏9⊐	⊏9⊐	⊏9⊐	⊏9⊐	⊏9⊐	⊏9⊐	⊏9⊐	⊏9⊐	⊏9⊐	⊏9⊐	⊏9⊐	⊏9⊐	⊏9⊐	⊏9⊐

ZIP CODE

⊏0⊐	⊏0⊐	⊏0⊐	⊏0⊐	⊏0⊐
⊏1⊐	⊏1⊐	⊏1⊐	⊏1⊐	⊏1⊐
⊏2⊐	⊏2⊐	⊏2⊐	⊏2⊐	⊏2⊐
⊏3⊐	⊏3⊐	⊏3⊐	⊏3⊐	⊏3⊐
⊏4⊐	⊏4⊐	⊏4⊐	⊏4⊐	⊏4⊐
⊏5⊐	⊏5⊐	⊏5⊐	⊏5⊐	⊏5⊐
⊏6⊐	⊏6⊐	⊏6⊐	⊏6⊐	⊏6⊐
⊏7⊐	⊏7⊐	⊏7⊐	⊏7⊐	⊏7⊐
⊏8⊐	⊏8⊐	⊏8⊐	⊏8⊐	⊏8⊐
⊏9⊐	⊏9⊐	⊏9⊐	⊏9⊐	⊏9⊐

FOLD HERE

FEED THIS DIRECTION

P4 0795-C1005-5 4 3 2

PreTest® Series

Published by
McGraw-Hill, Inc.
PreTest® Series.
Copyright© 1995 by
McGraw-Hill, Inc.
All rights reserved.
Printed in the U.S.A.

Side 2

CODE

ENTER THE
FIVE-DIGIT
CODE HERE

□0□	□0□	□0□	□0□	□0□
□1□	□1□	□1□	□1□	□1□
□2□	□2□	□2□	□2□	□2□
□3□	□3□	□3□	□3□	□3□
□4□	□4□	□4□	□4□	□4□
□5□	□5□	□5□	□5□	□5□
□6□	□6□	□6□	□6□	□6□
□7□	□7□	□7□	□7□	□7□
□8□	□8□	□8□	□8□	□8□
□9□	□9□	□9□	□9□	□9□

FEED THIS DIRECTION

15. □A□ □B□ □C□ □D□ □E□ □F□ □G□
□H□ □I□ □J□ □K□ □L□ □M□ □N□
□O□ □P□ □Q□ □R□ □S□ □T□ □U□

16. □A□ □B□ □C□ □D□ □E□ □F□ □G□
□H□ □I□ □J□ □K□ □L□ □M□ □N□
□O□ □P□ □Q□ □R□ □S□ □T□ □U□

17. □A□ □B□ □C□ □D□ □E□ □F□ □G□
□H□ □I□ □J□ □K□ □L□ □M□ □N□
□O□ □P□ □Q□ □R□ □S□ □T□ □U□

18. □A□ □B□ □C□ □D□ □E□ □F□ □G□
□H□ □I□ □J□ □K□ □L□ □M□ □N□
□O□ □P□ □Q□ □R□ □S□ □T□ □U□

19. □A□ □B□ □C□ □D□ □E□ □F□ □G□
□H□ □I□ □J□ □K□ □L□ □M□ □N□
□O□ □P□ □Q□ □R□ □S□ □T□ □U□

1. □A□ □B□ □C□ □D□ □E□ □F□ □G□
□H□ □I□ □J□ □K□ □L□ □M□ □N□
□O□ □P□ □Q□ □R□ □S□ □T□ □U□

2. □A□ □B□ □C□ □D□ □E□ □F□ □G□
□H□ □I□ □J□ □K□ □L□ □M□ □N□
□O□ □P□ □Q□ □R□ □S□ □T□ □U□

3. □A□ □B□ □C□ □D□ □E□ □F□ □G□
□H□ □I□ □J□ □K□ □L□ □M□ □N□
□O□ □P□ □Q□ □R□ □S□ □T□ □U□

4. □A□ □B□ □C□ □D□ □E□ □F□ □G□
□H□ □I□ □J□ □K□ □L□ □M□ □N□
□O□ □P□ □Q□ □R□ □S□ □T□ □U□

5. □A□ □B□ □C□ □D□ □E□ □F□ □G□
□H□ □I□ □J□ □K□ □L□ □M□ □N□
□O□ □P□ □Q□ □R□ □S□ □T□ □U□

6. □A□ □B□ □C□ □D□ □E□ □F□ □G□
□H□ □I□ □J□ □K□ □L□ □M□ □N□
□O□ □P□ □Q□ □R□ □S□ □T□ □U□

7. □A□ □B□ □C□ □D□ □E□ □F□ □G□
□H□ □I□ □J□ □K□ □L□ □M□ □N□
□O□ □P□ □Q□ □R□ □S□ □T□ □U□

8. □A□ □B□ □C□ □D□ □E□ □F□ □G□
□H□ □I□ □J□ □K□ □L□ □M□ □N□
□O□ □P□ □Q□ □R□ □S□ □T□ □U□

9. □A□ □B□ □C□ □D□ □E□ □F□ □G□
□H□ □I□ □J□ □K□ □L□ □M□ □N□
□O□ □P□ □Q□ □R□ □S□ □T□ □U□

10. □A□ □B□ □C□ □D□ □E□ □F□ □G□
□H□ □I□ □J□ □K□ □L□ □M□ □N□
□O□ □P□ □Q□ □R□ □S□ □T□ □U□

11. □A□ □B□ □C□ □D□ □E□ □F□ □G□
□H□ □I□ □J□ □K□ □L□ □M□ □N□
□O□ □P□ □Q□ □R□ □S□ □T□ □U□

12. □A□ □B□ □C□ □D□ □E□ □F□ □G□
□H□ □I□ □J□ □K□ □L□ □M□ □N□
□O□ □P□ □Q□ □R□ □S□ □T□ □U□

13. □A□ □B□ □C□ □D□ □E□ □F□ □G□
□H□ □I□ □J□ □K□ □L□ □M□ □N□
□O□ □P□ □Q□ □R□ □S□ □T□ □U□

14. □A□ □B□ □C□ □D□ □E□ □F□ □G□
□H□ □I□ □J□ □K□ □L□ □M□ □N□
□O□ □P□ □Q□ □R□ □S□ □T□ □U□

20. □A□ □B□ □C□ □D□ □E□ □F□ □G□
□H□ □I□ □J□ □K□ □L□ □M□ □N□
□O□ □P□ □Q□ □R□ □S□ □T□ □U□

21. □A□ □B□ □C□ □D□ □E□ □F□ □G□
□H□ □I□ □J□ □K□ □L□ □M□ □N□
□O□ □P□ □Q□ □R□ □S□ □T□ □U□

22. □A□ □B□ □C□ □D□ □E□ □F□ □G□
□H□ □I□ □J□ □K□ □L□ □M□ □N□
□O□ □P□ □Q□ □R□ □S□ □T□ □U□

23. □A□ □B□ □C□ □D□ □E□ □F□ □G□
□H□ □I□ □J□ □K□ □L□ □M□ □N□
□O□ □P□ □Q□ □R□ □S□ □T□ □U□

24. □A□ □B□ □C□ □D□ □E□ □F□ □G□
□H□ □I□ □J□ □K□ □L□ □M□ □N□
□O□ □P□ □Q□ □R□ □S□ □T□ □U□

25. □A□ □B□ □C□ □D□ □E□ □F□ □G□
□H□ □I□ □J□ □K□ □L□ □M□ □N□
□O□ □P□ □Q□ □R□ □S□ □T□ □U□

26. □A□ □B□ □C□ □D□ □E□ □F□ □G□
□H□ □I□ □J□ □K□ □L□ □M□ □N□
□O□ □P□ □Q□ □R□ □S□ □T□ □U□

27. □A□ □B□ □C□ □D□ □E□ □F□ □G□
□H□ □I□ □J□ □K□ □L□ □M□ □N□
□O□ □P□ □Q□ □R□ □S□ □T□ □U□

28. □A□ □B□ □C□ □D□ □E□ □F□ □G□
□H□ □I□ □J□ □K□ □L□ □M□ □N□
□O□ □P□ □Q□ □R□ □S□ □T□ □U□

29. □A□ □B□ □C□ □D□ □E□ □F□ □G□
□H□ □I□ □J□ □K□ □L□ □M□ □N□
□O□ □P□ □Q□ □R□ □S□ □T□ □U□

30. □A□ □B□ □C□ □D□ □E□ □F□ □G□
□H□ □I□ □J□ □K□ □L□ □M□ □N□
□O□ □P□ □Q□ □R□ □S□ □T□ □U□

31. □A□ □B□ □C□ □D□ □E□ □F□ □G□
□H□ □I□ □J□ □K□ □L□ □M□ □N□
□O□ □P□ □Q□ □R□ □S□ □T□ □U□

32. □A□ □B□ □C□ □D□ □E□ □F□ □G□
□H□ □I□ □J□ □K□ □L□ □M□ □N□
□O□ □P□ □Q□ □R□ □S□ □T□ □U□

33. □A□ □B□ □C□ □D□ □E□ □F□ □G□
□H□ □I□ □J□ □K□ □L□ □M□ □N□
□O□ □P□ □Q□ □R□ □S□ □T□ □U□

34. □A□ □B□ □C□ □D□ □E□ □F□ □G□
□H□ □I□ □J□ □K□ □L□ □M□ □N□
□O□ □P□ □Q□ □R□ □S□ □T□ □U□

35. □A□ □B□ □C□ □D□ □E□ □F□ □G□
□H□ □I□ □J□ □K□ □L□ □M□ □N□
□O□ □P□ □Q□ □R□ □S□ □T□ □U□

36. □A□ □B□ □C□ □D□ □E□ □F□ □G□
□H□ □I□ □J□ □K□ □L□ □M□ □N□
□O□ □P□ □Q□ □R□ □S□ □T□ □U□

37. □A□ □B□ □C□ □D□ □E□ □F□ □G□
□H□ □I□ □J□ □K□ □L□ □M□ □N□
□O□ □P□ □Q□ □R□ □S□ □T□ □U□

38. □A□ □B□ □C□ □D□ □E□ □F□ □G□
□H□ □I□ □J□ □K□ □L□ □M□ □N□
□O□ □P□ □Q□ □R□ □S□ □T□ □U□

39. □A□ □B□ □C□ □D□ □E□ □F□ □G□
□H□ □I□ □J□ □K□ □L□ □M□ □N□
□O□ □P□ □Q□ □R□ □S□ □T□ □U□

40. □A□ □B□ □C□ □D□ □E□ □F□ □G□ *FOLD*
□H□ □I□ □J□ □K□ □L□ □M□ □N□ *HERE*
□O□ □P□ □Q□ □R□ □S□ □T□ □U□

41. □A□ □B□ □C□ □D□ □E□ □F□ □G□
□H□ □I□ □J□ □K□ □L□ □M□ □N□
□O□ □P□ □Q□ □R□ □S□ □T□ □U□

42. □A□ □B□ □C□ □D□ □E□ □F□ □G□
□H□ □I□ □J□ □K□ □L□ □M□ □N□
□O□ □P□ □Q□ □R□ □S□ □T□ □U□

43. □A□ □B□ □C□ □D□ □E□ □F□ □G□
□H□ □I□ □J□ □K□ □L□ □M□ □N□
□O□ □P□ □Q□ □R□ □S□ □T□ □U□

44. □A□ □B□ □C□ □D□ □E□ □F□ □G□
□H□ □I□ □J□ □K□ □L□ □M□ □N□
□O□ □P□ □Q□ □R□ □S□ □T□ □U□

45. □A□ □B□ □C□ □D□ □E□ □F□ □G□
□H□ □I□ □J□ □K□ □L□ □M□ □N□
□O□ □P□ □Q□ □R□ □S□ □T□ □U□

46. □A□ □B□ □C□ □D□ □E□ □F□ □G□
□H□ □I□ □J□ □K□ □L□ □M□ □N□
□O□ □P□ □Q□ □R□ □S□ □T□ □U□

47. □A□ □B□ □C□ □D□ □E□ □F□ □G□ *FOLD*
□H□ □I□ □J□ □K□ □L□ □M□ □N□ *HERE*
□O□ □P□ □Q□ □R□ □S□ □T□ □U□

48. □A□ □B□ □C□ □D□ □E□ □F□ □G□
□H□ □I□ □J□ □K□ □L□ □M□ □N□
□O□ □P□ □Q□ □R□ □S□ □T□ □U□

49. □A□ □B□ □C□ □D□ □E□
50. □A□ □B□ □C□ □D□ □E□
51. □A□ □B□ □C□ □D□ □E□
52. □A□ □B□ □C□ □D□ □E□
53. □A□ □B□ □C□ □D□ □E□
54. □A□ □B□ □C□ □D□ □E□
55. □A□ □B□ □C□ □D□ □E□
56. □A□ □B□ □C□ □D□ □E□

Published by McGraw-Hill, Inc. PreTest® Series.
Copyright© 1995 by McGraw-Hill, Inc.
All Rights Reserved. Printed in the U.S.A.

Side 3

FEED THIS DIRECTION

57. ⊏A⊐ ⊏B⊐ ⊏C⊐ ⊏D⊐ ⊏E⊐
58. ⊏A⊐ ⊏B⊐ ⊏C⊐ ⊏D⊐ ⊏E⊐
59. ⊏A⊐ ⊏B⊐ ⊏C⊐ ⊏D⊐ ⊏E⊐
60. ⊏A⊐ ⊏B⊐ ⊏C⊐ ⊏D⊐ ⊏E⊐
61. ⊏A⊐ ⊏B⊐ ⊏C⊐ ⊏D⊐ ⊏E⊐
62. ⊏A⊐ ⊏B⊐ ⊏C⊐ ⊏D⊐ ⊏E⊐
63. ⊏A⊐ ⊏B⊐ ⊏C⊐ ⊏D⊐ ⊏E⊐
64. ⊏A⊐ ⊏B⊐ ⊏C⊐ ⊏D⊐ ⊏E⊐
65. ⊏A⊐ ⊏B⊐ ⊏C⊐ ⊏D⊐ ⊏E⊐
66. ⊏A⊐ ⊏B⊐ ⊏C⊐ ⊏D⊐ ⊏E⊐
67. ⊏A⊐ ⊏B⊐ ⊏C⊐ ⊏D⊐ ⊏E⊐
68. ⊏A⊐ ⊏B⊐ ⊏C⊐ ⊏D⊐ ⊏E⊐
69. ⊏A⊐ ⊏B⊐ ⊏C⊐ ⊏D⊐ ⊏E⊐
70. ⊏A⊐ ⊏B⊐ ⊏C⊐ ⊏D⊐ ⊏E⊐
71. ⊏A⊐ ⊏B⊐ ⊏C⊐ ⊏D⊐ ⊏E⊐
72. ⊏A⊐ ⊏B⊐ ⊏C⊐ ⊏D⊐ ⊏E⊐
73. ⊏A⊐ ⊏B⊐ ⊏C⊐ ⊏D⊐ ⊏E⊐
74. ⊏A⊐ ⊏B⊐ ⊏C⊐ ⊏D⊐ ⊏E⊐
75. ⊏A⊐ ⊏B⊐ ⊏C⊐ ⊏D⊐ ⊏E⊐
76. ⊏A⊐ ⊏B⊐ ⊏C⊐ ⊏D⊐ ⊏E⊐
77. ⊏A⊐ ⊏B⊐ ⊏C⊐ ⊏D⊐ ⊏E⊐
78. ⊏A⊐ ⊏B⊐ ⊏C⊐ ⊏D⊐ ⊏E⊐
79. ⊏A⊐ ⊏B⊐ ⊏C⊐ ⊏D⊐ ⊏E⊐
80. ⊏A⊐ ⊏B⊐ ⊏C⊐ ⊏D⊐ ⊏E⊐
81. ⊏A⊐ ⊏B⊐ ⊏C⊐ ⊏D⊐ ⊏E⊐
82. ⊏A⊐ ⊏B⊐ ⊏C⊐ ⊏D⊐ ⊏E⊐
83. ⊏A⊐ ⊏B⊐ ⊏C⊐ ⊏D⊐ ⊏E⊐
84. ⊏A⊐ ⊏B⊐ ⊏C⊐ ⊏D⊐ ⊏E⊐
85. ⊏A⊐ ⊏B⊐ ⊏C⊐ ⊏D⊐ ⊏E⊐
86. ⊏A⊐ ⊏B⊐ ⊏C⊐ ⊏D⊐ ⊏E⊐
87. ⊏A⊐ ⊏B⊐ ⊏C⊐ ⊏D⊐ ⊏E⊐
88. ⊏A⊐ ⊏B⊐ ⊏C⊐ ⊏D⊐ ⊏E⊐
89. ⊏A⊐ ⊏B⊐ ⊏C⊐ ⊏D⊐ ⊏E⊐
90. ⊏A⊐ ⊏B⊐ ⊏C⊐ ⊏D⊐ ⊏E⊐
91. ⊏A⊐ ⊏B⊐ ⊏C⊐ ⊏D⊐ ⊏E⊐
92. ⊏A⊐ ⊏B⊐ ⊏C⊐ ⊏D⊐ ⊏E⊐
93. ⊏A⊐ ⊏B⊐ ⊏C⊐ ⊏D⊐ ⊏E⊐
94. ⊏A⊐ ⊏B⊐ ⊏C⊐ ⊏D⊐ ⊏E⊐
95. ⊏A⊐ ⊏B⊐ ⊏C⊐ ⊏D⊐ ⊏E⊐
96. ⊏A⊐ ⊏B⊐ ⊏C⊐ ⊏D⊐ ⊏E⊐
97. ⊏A⊐ ⊏B⊐ ⊏C⊐ ⊏D⊐ ⊏E⊐
98. ⊏A⊐ ⊏B⊐ ⊏C⊐ ⊏D⊐ ⊏E⊐

99. ⊏A⊐ ⊏B⊐ ⊏C⊐ ⊏D⊐ ⊏E⊐
100. ⊏A⊐ ⊏B⊐ ⊏C⊐ ⊏D⊐ ⊏E⊐
101. ⊏A⊐ ⊏B⊐ ⊏C⊐ ⊏D⊐ ⊏E⊐
102. ⊏A⊐ ⊏B⊐ ⊏C⊐ ⊏D⊐ ⊏E⊐
103. ⊏A⊐ ⊏B⊐ ⊏C⊐ ⊏D⊐ ⊏E⊐
104. ⊏A⊐ ⊏B⊐ ⊏C⊐ ⊏D⊐ ⊏E⊐
105. ⊏A⊐ ⊏B⊐ ⊏C⊐ ⊏D⊐ ⊏E⊐
106. ⊏A⊐ ⊏B⊐ ⊏C⊐ ⊏D⊐ ⊏E⊐
107. ⊏A⊐ ⊏B⊐ ⊏C⊐ ⊏D⊐ ⊏E⊐
108. ⊏A⊐ ⊏B⊐ ⊏C⊐ ⊏D⊐ ⊏E⊐
109. ⊏A⊐ ⊏B⊐ ⊏C⊐ ⊏D⊐ ⊏E⊐
110. ⊏A⊐ ⊏B⊐ ⊏C⊐ ⊏D⊐ ⊏E⊐
111. ⊏A⊐ ⊏B⊐ ⊏C⊐ ⊏D⊐ ⊏E⊐
112. ⊏A⊐ ⊏B⊐ ⊏C⊐ ⊏D⊐ ⊏E⊐
113. ⊏A⊐ ⊏B⊐ ⊏C⊐ ⊏D⊐ ⊏E⊐
114. ⊏A⊐ ⊏B⊐ ⊏C⊐ ⊏D⊐ ⊏E⊐
115. ⊏A⊐ ⊏B⊐ ⊏C⊐ ⊏D⊐ ⊏E⊐
116. ⊏A⊐ ⊏B⊐ ⊏C⊐ ⊏D⊐ ⊏E⊐
117. ⊏A⊐ ⊏B⊐ ⊏C⊐ ⊏D⊐ ⊏E⊐
118. ⊏A⊐ ⊏B⊐ ⊏C⊐ ⊏D⊐ ⊏E⊐
119. ⊏A⊐ ⊏B⊐ ⊏C⊐ ⊏D⊐ ⊏E⊐
120. ⊏A⊐ ⊏B⊐ ⊏C⊐ ⊏D⊐ ⊏E⊐
121. ⊏A⊐ ⊏B⊐ ⊏C⊐ ⊏D⊐ ⊏E⊐
122. ⊏A⊐ ⊏B⊐ ⊏C⊐ ⊏D⊐ ⊏E⊐
123. ⊏A⊐ ⊏B⊐ ⊏C⊐ ⊏D⊐ ⊏E⊐
124. ⊏A⊐ ⊏B⊐ ⊏C⊐ ⊏D⊐ ⊏E⊐
125. ⊏A⊐ ⊏B⊐ ⊏C⊐ ⊏D⊐ ⊏E⊐
126. ⊏A⊐ ⊏B⊐ ⊏C⊐ ⊏D⊐ ⊏E⊐
127. ⊏A⊐ ⊏B⊐ ⊏C⊐ ⊏D⊐ ⊏E⊐
128. ⊏A⊐ ⊏B⊐ ⊏C⊐ ⊏D⊐ ⊏E⊐
129. ⊏A⊐ ⊏B⊐ ⊏C⊐ ⊏D⊐ ⊏E⊐
130. ⊏A⊐ ⊏B⊐ ⊏C⊐ ⊏D⊐ ⊏E⊐
131. ⊏A⊐ ⊏B⊐ ⊏C⊐ ⊏D⊐ ⊏E⊐
132. ⊏A⊐ ⊏B⊐ ⊏C⊐ ⊏D⊐ ⊏E⊐
133. ⊏A⊐ ⊏B⊐ ⊏C⊐ ⊏D⊐ ⊏E⊐
134. ⊏A⊐ ⊏B⊐ ⊏C⊐ ⊏D⊐ ⊏E⊐
135. ⊏A⊐ ⊏B⊐ ⊏C⊐ ⊏D⊐ ⊏E⊐
136. ⊏A⊐ ⊏B⊐ ⊏C⊐ ⊏D⊐ ⊏E⊐
137. ⊏A⊐ ⊏B⊐ ⊏C⊐ ⊏D⊐ ⊏E⊐
138. ⊏A⊐ ⊏B⊐ ⊏C⊐ ⊏D⊐ ⊏E⊐
139. ⊏A⊐ ⊏B⊐ ⊏C⊐ ⊏D⊐ ⊏E⊐
140. ⊏A⊐ ⊏B⊐ ⊏C⊐ ⊏D⊐ ⊏E⊐
141. ⊏A⊐ ⊏B⊐ ⊏C⊐ ⊏D⊐ ⊏E⊐
142. ⊏A⊐ ⊏B⊐ ⊏C⊐ ⊏D⊐ ⊏E⊐
143. ⊏A⊐ ⊏B⊐ ⊏C⊐ ⊏D⊐ ⊏E⊐
144. ⊏A⊐ ⊏B⊐ ⊏C⊐ ⊏D⊐ ⊏E⊐
145. ⊏A⊐ ⊏B⊐ ⊏C⊐ ⊏D⊐ ⊏E⊐
146. ⊏A⊐ ⊏B⊐ ⊏C⊐ ⊏D⊐ ⊏E⊐
147. ⊏A⊐ ⊏B⊐ ⊏C⊐ ⊏D⊐ ⊏E⊐
148. ⊏A⊐ ⊏B⊐ ⊏C⊐ ⊏D⊐ ⊏E⊐
149. ⊏A⊐ ⊏B⊐ ⊏C⊐ ⊏D⊐ ⊏E⊐
150. ⊏A⊐ ⊏B⊐ ⊏C⊐ ⊏D⊐ ⊏E⊐
151. ⊏A⊐ ⊏B⊐ ⊏C⊐ ⊏D⊐ ⊏E⊐
152. ⊏A⊐ ⊏B⊐ ⊏C⊐ ⊏D⊐ ⊏E⊐
153. ⊏A⊐ ⊏B⊐ ⊏C⊐ ⊏D⊐ ⊏E⊐
154. ⊏A⊐ ⊏B⊐ ⊏C⊐ ⊏D⊐ ⊏E⊐
155. ⊏A⊐ ⊏B⊐ ⊏C⊐ ⊏D⊐ ⊏E⊐
156. ⊏A⊐ ⊏B⊐ ⊏C⊐ ⊏D⊐ ⊏E⊐

157. ⊏A⊐ ⊏B⊐ ⊏C⊐ ⊏D⊐ ⊏E⊐
158. ⊏A⊐ ⊏B⊐ ⊏C⊐ ⊏D⊐ ⊏E⊐
159. ⊏A⊐ ⊏B⊐ ⊏C⊐ ⊏D⊐ ⊏E⊐
160. ⊏A⊐ ⊏B⊐ ⊏C⊐ ⊏D⊐ ⊏E⊐
161. ⊏A⊐ ⊏B⊐ ⊏C⊐ ⊏D⊐ ⊏E⊐
162. ⊏A⊐ ⊏B⊐ ⊏C⊐ ⊏D⊐ ⊏E⊐
163. ⊏A⊐ ⊏B⊐ ⊏C⊐ ⊏D⊐ ⊏E⊐
164. ⊏A⊐ ⊏B⊐ ⊏C⊐ ⊏D⊐ ⊏E⊐
165. ⊏A⊐ ⊏B⊐ ⊏C⊐ ⊏D⊐ ⊏E⊐
166. ⊏A⊐ ⊏B⊐ ⊏C⊐ ⊏D⊐ ⊏E⊐
167. ⊏A⊐ ⊏B⊐ ⊏C⊐ ⊏D⊐ ⊏E⊐
168. ⊏A⊐ ⊏B⊐ ⊏C⊐ ⊏D⊐ ⊏E⊐
169. ⊏A⊐ ⊏B⊐ ⊏C⊐ ⊏D⊐ ⊏E⊐
170. ⊏A⊐ ⊏B⊐ ⊏C⊐ ⊏D⊐ ⊏E⊐
171. ⊏A⊐ ⊏B⊐ ⊏C⊐ ⊏D⊐ ⊏E⊐
172. ⊏A⊐ ⊏B⊐ ⊏C⊐ ⊏D⊐ ⊏E⊐
173. ⊏A⊐ ⊏B⊐ ⊏C⊐ ⊏D⊐ ⊏E⊐
174. ⊏A⊐ ⊏B⊐ ⊏C⊐ ⊏D⊐ ⊏E⊐
175. ⊏A⊐ ⊏B⊐ ⊏C⊐ ⊏D⊐ ⊏E⊐
176. ⊏A⊐ ⊏B⊐ ⊏C⊐ ⊏D⊐ ⊏E⊐ *FOLD*
177. ⊏A⊐ ⊏B⊐ ⊏C⊐ ⊏D⊐ ⊏E⊐ *HERE*
178. ⊏A⊐ ⊏B⊐ ⊏C⊐ ⊏D⊐ ⊏E⊐
179. ⊏A⊐ ⊏B⊐ ⊏C⊐ ⊏D⊐ ⊏E⊐
180. ⊏A⊐ ⊏B⊐ ⊏C⊐ ⊏D⊐ ⊏E⊐
181. ⊏A⊐ ⊏B⊐ ⊏C⊐ ⊏D⊐ ⊏E⊐
182. ⊏A⊐ ⊏B⊐ ⊏C⊐ ⊏D⊐ ⊏E⊐
183. ⊏A⊐ ⊏B⊐ ⊏C⊐ ⊏D⊐ ⊏E⊐
184. ⊏A⊐ ⊏B⊐ ⊏C⊐ ⊏D⊐ ⊏E⊐
185. ⊏A⊐ ⊏B⊐ ⊏C⊐ ⊏D⊐ ⊏E⊐
186. ⊏A⊐ ⊏B⊐ ⊏C⊐ ⊏D⊐ ⊏E⊐
187. ⊏A⊐ ⊏B⊐ ⊏C⊐ ⊏D⊐ ⊏E⊐
188. ⊏A⊐ ⊏B⊐ ⊏C⊐ ⊏D⊐ ⊏E⊐
189. ⊏A⊐ ⊏B⊐ ⊏C⊐ ⊏D⊐ ⊏E⊐
190. ⊏A⊐ ⊏B⊐ ⊏C⊐ ⊏D⊐ ⊏E⊐
191. ⊏A⊐ ⊏B⊐ ⊏C⊐ ⊏D⊐ ⊏E⊐
192. ⊏A⊐ ⊏B⊐ ⊏C⊐ ⊏D⊐ ⊏E⊐
193. ⊏A⊐ ⊏B⊐ ⊏C⊐ ⊏D⊐ ⊏E⊐
194. ⊏A⊐ ⊏B⊐ ⊏C⊐ ⊏D⊐ ⊏E⊐
195. ⊏A⊐ ⊏B⊐ ⊏C⊐ ⊏D⊐ ⊏E⊐
196. ⊏A⊐ ⊏B⊐ ⊏C⊐ ⊏D⊐ ⊏E⊐
197. ⊏A⊐ ⊏B⊐ ⊏C⊐ ⊏D⊐ ⊏E⊐ *FOLD*
198. ⊏A⊐ ⊏B⊐ ⊏C⊐ ⊏D⊐ ⊏E⊐ *HERE*
199. ⊏A⊐ ⊏B⊐ ⊏C⊐ ⊏D⊐ ⊏E⊐
200. ⊏A⊐ ⊏B⊐ ⊏C⊐ ⊏D⊐ ⊏E⊐
201. ⊏A⊐ ⊏B⊐ ⊏C⊐ ⊏D⊐ ⊏E⊐
202. ⊏A⊐ ⊏B⊐ ⊏C⊐ ⊏D⊐ ⊏E⊐
203. ⊏A⊐ ⊏B⊐ ⊏C⊐ ⊏D⊐ ⊏E⊐
204. ⊏A⊐ ⊏B⊐ ⊏C⊐ ⊏D⊐ ⊏E⊐
205. ⊏A⊐ ⊏B⊐ ⊏C⊐ ⊏D⊐ ⊏E⊐
206. ⊏A⊐ ⊏B⊐ ⊏C⊐ ⊏D⊐ ⊏E⊐
207. ⊏A⊐ ⊏B⊐ ⊏C⊐ ⊏D⊐ ⊏E⊐
208. ⊏A⊐ ⊏B⊐ ⊏C⊐ ⊏D⊐ ⊏E⊐
209. ⊏A⊐ ⊏B⊐ ⊏C⊐ ⊏D⊐ ⊏E⊐
210. ⊏A⊐ ⊏B⊐ ⊏C⊐ ⊏D⊐ ⊏E⊐

Published by McGraw-Hill, Inc. PreTest® Series.
Copyright © 1995 by McGraw-Hill, Inc.
All Rights Reserved. Printed in the U.S.A.

Side 4

CODE

ENTER THE FIVE-DIGIT CODE HERE

Each code column has bubbles: 0 1 2 3 4 5 6 7 8 9 (five columns)

225. A B C D E F G / H I J K L M N / O P Q R S T U
226. A B C D E F G / H I J K L M N / O P Q R S T U
227. A B C D E F G / H I J K L M N / O P Q R S T U
228. A B C D E F G / H I J K L M N / O P Q R S T U
229. A B C D E F G / H I J K L M N / O P Q R S T U
230. A B C D E F G / H I J K L M N / O P Q R S T U
231. A B C D E F G / H I J K L M N / O P Q R S T U
232. A B C D E F G / H I J K L M N / O P Q R S T U
233. A B C D E F G / H I J K L M N / O P Q R S T U
234. A B C D E F G / H I J K L M N / O P Q R S T U
235. A B C D E F G / H I J K L M N / O P Q R S T U
236. A B C D E F G / H I J K L M N / O P Q R S T U
237. A B C D E F G / H I J K L M N / O P Q R S T U
238. A B C D E F G / H I J K L M N / O P Q R S T U
239. A B C D E F G / H I J K L M N / O P Q R S T U
240. A B C D E F G / H I J K L M N / O P Q R S T U
241. A B C D E F G / H I J K L M N / O P Q R S T U
242. A B C D E F G / H I J K L M N / O P Q R S T U
243. A B C D E F G / H I J K L M N / O P Q R S T U

244. A B C D E F G / H I J K L M N / O P Q R S T U
245. A B C D E F G / H I J K L M N / O P Q R S T U
246. A B C D E F G / H I J K L M N / O P Q R S T U
247. A B C D E F G / H I J K L M N / O P Q R S T U
248. A B C D E F G / H I J K L M N / O P Q R S T U
249. A B C D E F G / H I J K L M N / O P Q R S T U
250. A B C D E F G / H I J K L M N / O P Q R S T U *FOLD HERE*
251. A B C D E F G / H I J K L M N / O P Q R S T U
252. A B C D E F G / H I J K L M N / O P Q R S T U
253. A B C D E F G / H I J K L M N / O P Q R S T U
254. A B C D E F G / H I J K L M N / O P Q R S T U
255. A B C D E F G / H I J K L M N / O P Q R S T U
256. A B C D E F G / H I J K L M N / O P Q R S T U
257. A B C D E F G / H I J K L M N / O P Q R S T U *FOLD HERE*
258. A B C D E F G / H I J K L M N / O P Q R S T U

211. A B C D E F G / H I J K L M N / O P Q R S T U
212. A B C D E F G / H I J K L M N / O P Q R S T U
213. A B C D E F G / H I J K L M N / O P Q R S T U
214. A B C D E F G / H I J K L M N / O P Q R S T U
215. A B C D E F G / H I J K L M N / O P Q R S T U
216. A B C D E F G / H I J K L M N / O P Q R S T U
217. A B C D E F G / H I J K L M N / O P Q R S T U
218. A B C D E F G / H I J K L M N / O P Q R S T U
219. A B C D E F G / H I J K L M N / O P Q R S T U
220. A B C D E F G / H I J K L M N / O P Q R S T U
221. A B C D E F G / H I J K L M N / O P Q R S T U
222. A B C D E F G / H I J K L M N / O P Q R S T U
223. A B C D E F G / H I J K L M N / O P Q R S T U
224. A B C D E F G / H I J K L M N / O P Q R S T U

259. A B C D E
260. A B C D E
261. A B C D E
262. A B C D E
263. A B C D E
264. A B C D E
265. A B C D E
266. A B C D E

FEED THIS DIRECTION

Published by McGraw-Hill, Inc. PreTest® Series.
Copyright © 1995 by McGraw-Hill, Inc.
All Rights Reserved. Printed in the U.S.A.

Side 5

CODE

ENTER THE FIVE-DIGIT CODE HERE →

⊏0⊐	⊏0⊐	⊏0⊐	⊏0⊐	⊏0⊐
⊏1⊐	⊏1⊐	⊏1⊐	⊏1⊐	⊏1⊐
⊏2⊐	⊏2⊐	⊏2⊐	⊏2⊐	⊏2⊐
⊏3⊐	⊏3⊐	⊏3⊐	⊏3⊐	⊏3⊐
⊏4⊐	⊏4⊐	⊏4⊐	⊏4⊐	⊏4⊐
⊏5⊐	⊏5⊐	⊏5⊐	⊏5⊐	⊏5⊐
⊏6⊐	⊏6⊐	⊏6⊐	⊏6⊐	⊏6⊐
⊏7⊐	⊏7⊐	⊏7⊐	⊏7⊐	⊏7⊐
⊏8⊐	⊏8⊐	⊏8⊐	⊏8⊐	⊏8⊐
⊏9⊐	⊏9⊐	⊏9⊐	⊏9⊐	⊏9⊐

FEED THIS DIRECTION

267. ⊏A⊐ ⊏B⊐ ⊏C⊐ ⊏D⊐ ⊏E⊐
268. ⊏A⊐ ⊏B⊐ ⊏C⊐ ⊏D⊐ ⊏E⊐
269. ⊏A⊐ ⊏B⊐ ⊏C⊐ ⊏D⊐ ⊏E⊐
270. ⊏A⊐ ⊏B⊐ ⊏C⊐ ⊏D⊐ ⊏E⊐
271. ⊏A⊐ ⊏B⊐ ⊏C⊐ ⊏D⊐ ⊏E⊐
272. ⊏A⊐ ⊏B⊐ ⊏C⊐ ⊏D⊐ ⊏E⊐
273. ⊏A⊐ ⊏B⊐ ⊏C⊐ ⊏D⊐ ⊏E⊐
274. ⊏A⊐ ⊏B⊐ ⊏C⊐ ⊏D⊐ ⊏E⊐
275. ⊏A⊐ ⊏B⊐ ⊏C⊐ ⊏D⊐ ⊏E⊐
276. ⊏A⊐ ⊏B⊐ ⊏C⊐ ⊏D⊐ ⊏E⊐
277. ⊏A⊐ ⊏B⊐ ⊏C⊐ ⊏D⊐ ⊏E⊐
278. ⊏A⊐ ⊏B⊐ ⊏C⊐ ⊏D⊐ ⊏E⊐
279. ⊏A⊐ ⊏B⊐ ⊏C⊐ ⊏D⊐ ⊏E⊐
280. ⊏A⊐ ⊏B⊐ ⊏C⊐ ⊏D⊐ ⊏E⊐
281. ⊏A⊐ ⊏B⊐ ⊏C⊐ ⊏D⊐ ⊏E⊐
282. ⊏A⊐ ⊏B⊐ ⊏C⊐ ⊏D⊐ ⊏E⊐
283. ⊏A⊐ ⊏B⊐ ⊏C⊐ ⊏D⊐ ⊏E⊐
284. ⊏A⊐ ⊏B⊐ ⊏C⊐ ⊏D⊐ ⊏E⊐
285. ⊏A⊐ ⊏B⊐ ⊏C⊐ ⊏D⊐ ⊏E⊐
286. ⊏A⊐ ⊏B⊐ ⊏C⊐ ⊏D⊐ ⊏E⊐
287. ⊏A⊐ ⊏B⊐ ⊏C⊐ ⊏D⊐ ⊏E⊐
288. ⊏A⊐ ⊏B⊐ ⊏C⊐ ⊏D⊐ ⊏E⊐
289. ⊏A⊐ ⊏B⊐ ⊏C⊐ ⊏D⊐ ⊏E⊐
290. ⊏A⊐ ⊏B⊐ ⊏C⊐ ⊏D⊐ ⊏E⊐
291. ⊏A⊐ ⊏B⊐ ⊏C⊐ ⊏D⊐ ⊏E⊐
292. ⊏A⊐ ⊏B⊐ ⊏C⊐ ⊏D⊐ ⊏E⊐
293. ⊏A⊐ ⊏B⊐ ⊏C⊐ ⊏D⊐ ⊏E⊐
294. ⊏A⊐ ⊏B⊐ ⊏C⊐ ⊏D⊐ ⊏E⊐
295. ⊏A⊐ ⊏B⊐ ⊏C⊐ ⊏D⊐ ⊏E⊐
296. ⊏A⊐ ⊏B⊐ ⊏C⊐ ⊏D⊐ ⊏E⊐
297. ⊏A⊐ ⊏B⊐ ⊏C⊐ ⊏D⊐ ⊏E⊐
298. ⊏A⊐ ⊏B⊐ ⊏C⊐ ⊏D⊐ ⊏E⊐
299. ⊏A⊐ ⊏B⊐ ⊏C⊐ ⊏D⊐ ⊏E⊐
300. ⊏A⊐ ⊏B⊐ ⊏C⊐ ⊏D⊐ ⊏E⊐
301. ⊏A⊐ ⊏B⊐ ⊏C⊐ ⊏D⊐ ⊏E⊐
302. ⊏A⊐ ⊏B⊐ ⊏C⊐ ⊏D⊐ ⊏E⊐
303. ⊏A⊐ ⊏B⊐ ⊏C⊐ ⊏D⊐ ⊏E⊐
304. ⊏A⊐ ⊏B⊐ ⊏C⊐ ⊏D⊐ ⊏E⊐
305. ⊏A⊐ ⊏B⊐ ⊏C⊐ ⊏D⊐ ⊏E⊐
306. ⊏A⊐ ⊏B⊐ ⊏C⊐ ⊏D⊐ ⊏E⊐
307. ⊏A⊐ ⊏B⊐ ⊏C⊐ ⊏D⊐ ⊏E⊐
308. ⊏A⊐ ⊏B⊐ ⊏C⊐ ⊏D⊐ ⊏E⊐

309. ⊏A⊐ ⊏B⊐ ⊏C⊐ ⊏D⊐ ⊏E⊐
310. ⊏A⊐ ⊏B⊐ ⊏C⊐ ⊏D⊐ ⊏E⊐
311. ⊏A⊐ ⊏B⊐ ⊏C⊐ ⊏D⊐ ⊏E⊐
312. ⊏A⊐ ⊏B⊐ ⊏C⊐ ⊏D⊐ ⊏E⊐
313. ⊏A⊐ ⊏B⊐ ⊏C⊐ ⊏D⊐ ⊏E⊐
314. ⊏A⊐ ⊏B⊐ ⊏C⊐ ⊏D⊐ ⊏E⊐
315. ⊏A⊐ ⊏B⊐ ⊏C⊐ ⊏D⊐ ⊏E⊐
316. ⊏A⊐ ⊏B⊐ ⊏C⊐ ⊏D⊐ ⊏E⊐
317. ⊏A⊐ ⊏B⊐ ⊏C⊐ ⊏D⊐ ⊏E⊐
318. ⊏A⊐ ⊏B⊐ ⊏C⊐ ⊏D⊐ ⊏E⊐
319. ⊏A⊐ ⊏B⊐ ⊏C⊐ ⊏D⊐ ⊏E⊐
320. ⊏A⊐ ⊏B⊐ ⊏C⊐ ⊏D⊐ ⊏E⊐
321. ⊏A⊐ ⊏B⊐ ⊏C⊐ ⊏D⊐ ⊏E⊐
322. ⊏A⊐ ⊏B⊐ ⊏C⊐ ⊏D⊐ ⊏E⊐
323. ⊏A⊐ ⊏B⊐ ⊏C⊐ ⊏D⊐ ⊏E⊐
324. ⊏A⊐ ⊏B⊐ ⊏C⊐ ⊏D⊐ ⊏E⊐
325. ⊏A⊐ ⊏B⊐ ⊏C⊐ ⊏D⊐ ⊏E⊐
326. ⊏A⊐ ⊏B⊐ ⊏C⊐ ⊏D⊐ ⊏E⊐
327. ⊏A⊐ ⊏B⊐ ⊏C⊐ ⊏D⊐ ⊏E⊐
328. ⊏A⊐ ⊏B⊐ ⊏C⊐ ⊏D⊐ ⊏E⊐
329. ⊏A⊐ ⊏B⊐ ⊏C⊐ ⊏D⊐ ⊏E⊐
330. ⊏A⊐ ⊏B⊐ ⊏C⊐ ⊏D⊐ ⊏E⊐
331. ⊏A⊐ ⊏B⊐ ⊏C⊐ ⊏D⊐ ⊏E⊐
332. ⊏A⊐ ⊏B⊐ ⊏C⊐ ⊏D⊐ ⊏E⊐
333. ⊏A⊐ ⊏B⊐ ⊏C⊐ ⊏D⊐ ⊏E⊐
334. ⊏A⊐ ⊏B⊐ ⊏C⊐ ⊏D⊐ ⊏E⊐
335. ⊏A⊐ ⊏B⊐ ⊏C⊐ ⊏D⊐ ⊏E⊐
336. ⊏A⊐ ⊏B⊐ ⊏C⊐ ⊏D⊐ ⊏E⊐
337. ⊏A⊐ ⊏B⊐ ⊏C⊐ ⊏D⊐ ⊏E⊐
338. ⊏A⊐ ⊏B⊐ ⊏C⊐ ⊏D⊐ ⊏E⊐
339. ⊏A⊐ ⊏B⊐ ⊏C⊐ ⊏D⊐ ⊏E⊐
340. ⊏A⊐ ⊏B⊐ ⊏C⊐ ⊏D⊐ ⊏E⊐
341. ⊏A⊐ ⊏B⊐ ⊏C⊐ ⊏D⊐ ⊏E⊐
342. ⊏A⊐ ⊏B⊐ ⊏C⊐ ⊏D⊐ ⊏E⊐
343. ⊏A⊐ ⊏B⊐ ⊏C⊐ ⊏D⊐ ⊏E⊐
344. ⊏A⊐ ⊏B⊐ ⊏C⊐ ⊏D⊐ ⊏E⊐
345. ⊏A⊐ ⊏B⊐ ⊏C⊐ ⊏D⊐ ⊏E⊐
346. ⊏A⊐ ⊏B⊐ ⊏C⊐ ⊏D⊐ ⊏E⊐
347. ⊏A⊐ ⊏B⊐ ⊏C⊐ ⊏D⊐ ⊏E⊐
348. ⊏A⊐ ⊏B⊐ ⊏C⊐ ⊏D⊐ ⊏E⊐
349. ⊏A⊐ ⊏B⊐ ⊏C⊐ ⊏D⊐ ⊏E⊐
350. ⊏A⊐ ⊏B⊐ ⊏C⊐ ⊏D⊐ ⊏E⊐
351. ⊏A⊐ ⊏B⊐ ⊏C⊐ ⊏D⊐ ⊏E⊐
352. ⊏A⊐ ⊏B⊐ ⊏C⊐ ⊏D⊐ ⊏E⊐
353. ⊏A⊐ ⊏B⊐ ⊏C⊐ ⊏D⊐ ⊏E⊐
354. ⊏A⊐ ⊏B⊐ ⊏C⊐ ⊏D⊐ ⊏E⊐
355. ⊏A⊐ ⊏B⊐ ⊏C⊐ ⊏D⊐ ⊏E⊐
356. ⊏A⊐ ⊏B⊐ ⊏C⊐ ⊏D⊐ ⊏E⊐
357. ⊏A⊐ ⊏B⊐ ⊏C⊐ ⊏D⊐ ⊏E⊐
358. ⊏A⊐ ⊏B⊐ ⊏C⊐ ⊏D⊐ ⊏E⊐
359. ⊏A⊐ ⊏B⊐ ⊏C⊐ ⊏D⊐ ⊏E⊐
360. ⊏A⊐ ⊏B⊐ ⊏C⊐ ⊏D⊐ ⊏E⊐
361. ⊏A⊐ ⊏B⊐ ⊏C⊐ ⊏D⊐ ⊏E⊐
362. ⊏A⊐ ⊏B⊐ ⊏C⊐ ⊏D⊐ ⊏E⊐
363. ⊏A⊐ ⊏B⊐ ⊏C⊐ ⊏D⊐ ⊏E⊐
364. ⊏A⊐ ⊏B⊐ ⊏C⊐ ⊏D⊐ ⊏E⊐
365. ⊏A⊐ ⊏B⊐ ⊏C⊐ ⊏D⊐ ⊏E⊐
366. ⊏A⊐ ⊏B⊐ ⊏C⊐ ⊏D⊐ ⊏E⊐

367. ⊏A⊐ ⊏B⊐ ⊏C⊐ ⊏D⊐ ⊏E⊐
368. ⊏A⊐ ⊏B⊐ ⊏C⊐ ⊏D⊐ ⊏E⊐
369. ⊏A⊐ ⊏B⊐ ⊏C⊐ ⊏D⊐ ⊏E⊐
370. ⊏A⊐ ⊏B⊐ ⊏C⊐ ⊏D⊐ ⊏E⊐
371. ⊏A⊐ ⊏B⊐ ⊏C⊐ ⊏D⊐ ⊏E⊐
372. ⊏A⊐ ⊏B⊐ ⊏C⊐ ⊏D⊐ ⊏E⊐
373. ⊏A⊐ ⊏B⊐ ⊏C⊐ ⊏D⊐ ⊏E⊐
374. ⊏A⊐ ⊏B⊐ ⊏C⊐ ⊏D⊐ ⊏E⊐
375. ⊏A⊐ ⊏B⊐ ⊏C⊐ ⊏D⊐ ⊏E⊐
376. ⊏A⊐ ⊏B⊐ ⊏C⊐ ⊏D⊐ ⊏E⊐
377. ⊏A⊐ ⊏B⊐ ⊏C⊐ ⊏D⊐ ⊏E⊐
378. ⊏A⊐ ⊏B⊐ ⊏C⊐ ⊏D⊐ ⊏E⊐
379. ⊏A⊐ ⊏B⊐ ⊏C⊐ ⊏D⊐ ⊏E⊐
380. ⊏A⊐ ⊏B⊐ ⊏C⊐ ⊏D⊐ ⊏E⊐
381. ⊏A⊐ ⊏B⊐ ⊏C⊐ ⊏D⊐ ⊏E⊐
382. ⊏A⊐ ⊏B⊐ ⊏C⊐ ⊏D⊐ ⊏E⊐
383. ⊏A⊐ ⊏B⊐ ⊏C⊐ ⊏D⊐ ⊏E⊐
384. ⊏A⊐ ⊏B⊐ ⊏C⊐ ⊏D⊐ ⊏E⊐
385. ⊏A⊐ ⊏B⊐ ⊏C⊐ ⊏D⊐ ⊏E⊐
386. ⊏A⊐ ⊏B⊐ ⊏C⊐ ⊏D⊐ ⊏E⊐   *FOLD*
387. ⊏A⊐ ⊏B⊐ ⊏C⊐ ⊏D⊐ ⊏E⊐   *HERE*
388. ⊏A⊐ ⊏B⊐ ⊏C⊐ ⊏D⊐ ⊏E⊐
389. ⊏A⊐ ⊏B⊐ ⊏C⊐ ⊏D⊐ ⊏E⊐
390. ⊏A⊐ ⊏B⊐ ⊏C⊐ ⊏D⊐ ⊏E⊐
391. ⊏A⊐ ⊏B⊐ ⊏C⊐ ⊏D⊐ ⊏E⊐
392. ⊏A⊐ ⊏B⊐ ⊏C⊐ ⊏D⊐ ⊏E⊐
393. ⊏A⊐ ⊏B⊐ ⊏C⊐ ⊏D⊐ ⊏E⊐
394. ⊏A⊐ ⊏B⊐ ⊏C⊐ ⊏D⊐ ⊏E⊐
395. ⊏A⊐ ⊏B⊐ ⊏C⊐ ⊏D⊐ ⊏E⊐
396. ⊏A⊐ ⊏B⊐ ⊏C⊐ ⊏D⊐ ⊏E⊐
397. ⊏A⊐ ⊏B⊐ ⊏C⊐ ⊏D⊐ ⊏E⊐
398. ⊏A⊐ ⊏B⊐ ⊏C⊐ ⊏D⊐ ⊏E⊐
399. ⊏A⊐ ⊏B⊐ ⊏C⊐ ⊏D⊐ ⊏E⊐
400. ⊏A⊐ ⊏B⊐ ⊏C⊐ ⊏D⊐ ⊏E⊐
401. ⊏A⊐ ⊏B⊐ ⊏C⊐ ⊏D⊐ ⊏E⊐
402. ⊏A⊐ ⊏B⊐ ⊏C⊐ ⊏D⊐ ⊏E⊐
403. ⊏A⊐ ⊏B⊐ ⊏C⊐ ⊏D⊐ ⊏E⊐
404. ⊏A⊐ ⊏B⊐ ⊏C⊐ ⊏D⊐ ⊏E⊐
405. ⊏A⊐ ⊏B⊐ ⊏C⊐ ⊏D⊐ ⊏E⊐
406. ⊏A⊐ ⊏B⊐ ⊏C⊐ ⊏D⊐ ⊏E⊐
407. ⊏A⊐ ⊏B⊐ ⊏C⊐ ⊏D⊐ ⊏E⊐   *FOLD*
408. ⊏A⊐ ⊏B⊐ ⊏C⊐ ⊏D⊐ ⊏E⊐   *HERE*
409. ⊏A⊐ ⊏B⊐ ⊏C⊐ ⊏D⊐ ⊏E⊐
410. ⊏A⊐ ⊏B⊐ ⊏C⊐ ⊏D⊐ ⊏E⊐
411. ⊏A⊐ ⊏B⊐ ⊏C⊐ ⊏D⊐ ⊏E⊐
412. ⊏A⊐ ⊏B⊐ ⊏C⊐ ⊏D⊐ ⊏E⊐
413. ⊏A⊐ ⊏B⊐ ⊏C⊐ ⊏D⊐ ⊏E⊐
414. ⊏A⊐ ⊏B⊐ ⊏C⊐ ⊏D⊐ ⊏E⊐
415. ⊏A⊐ ⊏B⊐ ⊏C⊐ ⊏D⊐ ⊏E⊐
416. ⊏A⊐ ⊏B⊐ ⊏C⊐ ⊏D⊐ ⊏E⊐
417. ⊏A⊐ ⊏B⊐ ⊏C⊐ ⊏D⊐ ⊏E⊐
418. ⊏A⊐ ⊏B⊐ ⊏C⊐ ⊏D⊐ ⊏E⊐
419. ⊏A⊐ ⊏B⊐ ⊏C⊐ ⊏D⊐ ⊏E⊐
420. ⊏A⊐ ⊏B⊐ ⊏C⊐ ⊏D⊐ ⊏E⊐